Research Methods for Business and Social Science Students
(Second Edition)

Thank you for choosing a SAGE product!
If you have any comment, observation or feedback,
I would like to personally hear from you.

Please write to me at **contactceo@sagepub.in**

Vivek Mehra, Managing Director and CEO, SAGE India.

Bulk Sales

SAGE India offers special discounts
for purchase of books in bulk.
We also make available special imprints
and excerpts from our books on demand.

For orders and enquiries, write to us at

Marketing Department
SAGE Publications India Pvt Ltd
B1/I-1, Mohan Cooperative Industrial Area
Mathura Road, Post Bag 7
New Delhi 110044, India

E-mail us at **marketing@sagepub.in**

Get to know more about SAGE

Be invited to SAGE events, get on our mailing list.
Write today to **marketing@sagepub.in**

This book is also available as an e-book.

Research Methods for Business and Social Science Students
(Second Edition)

John Adams
Hafiz T. A. Khan
Robert Raeside

Los Angeles | London | New Delhi
Singapore | Washington DC | Melbourne

Copyright © John Adams, Hafiz T. A. Khan and Robert Raeside, 2014

All rights reserved. No part of this book may be reproduced or utilised in any form or by any means, electronic or mechanical, including photocopying, recording or by any information storage or retrieval system, without permission in writing from the publisher.

First published in 2007
This second edition published in 2014 by

SAGE Publications India Pvt Ltd
B1/I-1 Mohan Cooperative Industrial Area
Mathura Road, New Delhi 110 044, India
www.sagepub.in

SAGE Publications Inc
2455 Teller Road
Thousand Oaks, California 91320, USA

SAGE Publications Ltd
1 Oliver's Yard, 55 City Road
London EC1Y 1SP, United Kingdom

SAGE Publications Asia-Pacific Pte Ltd
3 Church Street
#10-04 Samsung Hub
Singapore 049483

Published by Vivek Mehra for SAGE Publications India Pvt Ltd, typeset in 11/13 Minion Pro by RECTO Graphics, Delhi and printed at Sai Print-o-Pack, New Delhi.

Library of Congress Cataloging-in-Publication Data

Adams, John.
 Research methods for business and social science students / John Adams, Hafiz T. A. Khan, Robert Raeside.—Second edition.
  pages cm
 Revised edition of: Research methods for graduate business and social science students / John Adams ... [et al.], published in 2007.
 Includes bibliographical references.
 1. Social sciences—Research. 2. Social sciences—Methodology. I. Khan, Hafiz T. A. II. Raeside, R. (Robert) III. Research methods for graduate business and social science students. IV. Title.
  H62.A5134   001.4'2—dc23   2014   2013049964

ISBN: 978-81-321-1366-9 (PB)

The SAGE Team: Sachin Sharma, Vandana Gupta, Rajib Chatterjee and Dally Verghese

Contents

Preface   xi

Acknowledgements   xiii

Chapter 1: Introduction to Research   1
- 1.1 Introduction   1
- 1.2 What Is Research?   1
- 1.3 Why Is Research Conducted?   3
- 1.4 Who Does Research?   3
- 1.5 How Is Research Conducted?   4
- 1.6 Business and Social Science Research Methods   4

Chapter 2: Research Methodology   5
- 2.1 Introduction   5
- 2.2 Research Method versus Research Methodology   5
- 2.3 Approaches to Business and Social Research   6
- 2.4 Justifying the Scientific Method   9
- 2.5 Research Ethics   21
- 2.6 Exercises   23
- 2.7 References   24

Chapter 3: The Research Cycle   26
- 3.1 Introduction   26
- 3.2 The Research Cycle   26
- 3.3 Problems with the Research Process   30
- 3.4 Exercises   32
- 3.5 References   32

Chapter 4: Literature Review and Critical Reading   33
- 4.1 Introduction   33
- 4.2 The Importance of a Literature Review   34
- 4.3 What Should the Literature Review Do?   38
- 4.4 Types of Literature Review   42

4.5	Some General Points in Literature Reviewing	43
4.6	Obtaining Literature Sources	44
4.7	Searching the Literature	45
4.8	Assessing the Quality of Literature	48
4.9	An Example of a Literature Review	49
4.10	Critical Evaluation	58
4.11	Critical Analysis	59
4.12	Critical Reading	59
4.13	Critical Thinking	60
4.14	Critical Questions	60
4.15	Critical Reviews	62
4.16	Exercise	63

Chapter 5: Sampling — 64

5.1	Introduction	64
5.2	Classification of Research Designs	65
5.3	Sources of Data	70
5.4	Types of Data and Measurement	70
5.5	Methods of Data Collection	72
5.6	Sampling Techniques	72
5.7	Representative Sampling Plans	73
5.8	Sample Size Determination	77
5.9	Test of Significance for Population Mean	78
5.10	Test of Significance for Population Proportion	79
5.11	Key Statistical Concepts	82
5.12	Some Problems with Random Sample Surveys	86
5.13	The Normal Distribution	90
5.14	Exercise	91
5.15	Reference	91

Chapter 6: Primary Data Collection — 92

6.1	Introduction	92
6.2	Observation	92
6.3	Experimentation	95
6.4	Surveys	96
6.5	Interviews	97
6.6	Diary Methods	97
6.7	Case Studies	98
6.8	Data Storage	99
6.9	Triangulation	99
6.10	Exercises	102
6.11	References	103
6.12	Websites	103

Chapter 7: Secondary Data Collection — 104
 7.1 Introduction — 104
 7.2 Web Search Skills — 106
 7.3 Exercises — 116
 7.4 References — 117

Chapter 8: Surveys — 118
 8.1 Introduction — 118
 8.2 Design — 119
 8.3 Questions — 123
 8.4 Pilot Survey — 127
 8.5 Administering the Survey — 128
 8.6 Ensuring High Response Rates — 130
 8.7 Missing Information — 132
 8.8 Coding and Data Input — 133
 8.9 Guidelines — 135
 8.10 Social Networks — 136
 8.11 Exercise — 141
 8.12 References — 141
 8.13 Websites — 142

Chapter 9: Interviews and Focus Groups — 143
 9.1 Introduction — 143
 9.2 Why Do Interviews? — 143
 9.3 General Guidelines for Interviewing — 145
 9.4 Bias and Errors — 148
 9.5 Telephone Interviews — 149
 9.6 Group/Focus Group Interviews — 150
 9.7 Reference — 151

Chapter 10: Qualitative Data Analysis — 152
 10.1 Introduction — 152
 10.2 Preparation — 153
 10.3 Content Analysis — 159
 10.4 Summarising — 164
 10.5 Grounded Theory — 167
 10.6 References — 168

Chapter 11: Descriptive Quantitative Analysis — 169
 11.1 Introduction — 169
 11.2 Descriptive Statistics — 171
 11.3 Are There Significant Differences? — 185
 11.4 Comparing Two Groups — 189
 11.5 Comparing More than Two Groups — 191

11.6	The Association between Categorical Variables	193
11.7	Summary of Test Procedures	195
11.8	Exercises	195
11.9	References	198

Chapter 12: Correlation and Regression — 199

12.1	Introduction	199
12.2	Correlation	199
12.3	Regression	202
12.4	Diagnostics	205
12.5	Multiple Regression	207
12.6	Modelling	213
12.7	Exercises	219
12.8	Reference	220

Chapter 13: Advanced Statistical Analysis — 221

13.1	Introduction	221
13.2	Factor Analysis	222
13.3	Logistic Regression	235
13.4	Exercises	243
13.5	References	244

Chapter 14: Tests of Measurement and Quality — 245

14.1	Introduction	245
14.2	Reliability	245
14.3	Validity	247
14.4	Generalisability	252
14.5	Exercises	253
14.6	Websites	253

Chapter 15: Conducting Your Research — 254

15.1	Introduction	254
15.2	Selecting the Topic	254
15.3	Guide to Supervision	258
15.4	Undertaking Your Research	259
15.5	Research Proposal Example	263
15.6	Reference	271

Chapter 16: Writing and Presenting the Dissertation — 272

16.1	Introduction	272
16.2	The Dissertation	272
16.3	Dissertation Objectives	273
16.4	What Should a Dissertation Look Like?	274
16.5	Presenting the Dissertation	275

Appendices **281**
    Appendix I: Confidentiality in Use of Data Provided by Third Parties   281
    Appendix II: Declaration   282
    Appendix III: Specimen Title Page (Front Cover)   282
    Appendix IV: Specimen Title Page (Inside Page)   283
    Appendix V: Multiple-choice Self-test   283

Bibliography and Further Reading   288

About the Authors   289

Preface

We were inspired by many students, colleagues and friends to compile these chapters and to produce a second edition of the book for business and social science students. The first edition was titled *Research Methods for Graduate Business and Social Science Students*; however, feedback has revealed that the book is widely used by both undergraduate and postgraduate students. Therefore, we decided to add more information and change the title to better reflect the wide range of users of the book. Our sincere thanks goes to them. As students of business or social science and students from other disciplines, we hope you will find it extremely useful. We have illustrated as much information as possible, and tried to facilitate ease of understanding. As part of this we have deliberately attempted to 'de-jargonise' the book and to present the material in as practical a manner as possible. In the space available we could not cover every topic but hope that the book will be sufficiently comprehensive. References to additional reading have been given and will hopefully overcome limitations arising from brevity.

Welcome to the subject of Research Methods. This will provide preparation before embarking on your own research, which will probably be a dissertation as part of your degree. Two very fundamental aims of Research Methods are:

1. To enable you to acquire knowledge and skills in the field of research methods
2. To prepare you to undertake research on your own applying the knowledge and skills of research methods on a research topic relevant to your area of study

The book is divided into sixteen chapters. In Chapter 1 the general concepts in relation to research are introduced. Broad research issues and theoretical concepts critical to research and research methods are the subject of Chapter 2. The importance of research ethics is also outlined in this chapter. In Chapter 3 the formulation of research along with the research process is discussed. Then in Chapter 4, we move onto an aspect of research that is often not treated with the importance it deserves but is fundamental to good research and to the synthesis and creation of knowledge—literature analysis and critical reading. In Chapter 5 an aspect of research is presented in terms of research design, that is, *how* to plan a research project and *how* to affect its implementation. This is, in many ways, the most important part of undertaking a research project. In Chapter 6 the concentration is on primary data collection for both qualitative and quantitative research. A detailed discussion on secondary sources of information is contained in Chapter 7. More detail on surveys is the subject of Chapter 8 and an indication of the important parts of design, questionnaires and data management is presented. The interview is also an important part of survey research and this is discussed in Chapter 9.

Chapters 10 and 11 deal with a number of research techniques covering both qualitative and quantitative research methods and how these methods are practically used to understand the real world. In this part you will find that the distinction between these two, in practice, is often fuzzy and real-world research often requires inputs from both approaches.

An overview of both elementary and advanced statistical analyses is given in Chapters 12 and 13 in order to give an understanding of statistical methods and their applications. For many students these chapters may be omitted; although we are strong advocates of quantitative approaches, we realise that not all share our passion and in many cases these methods may not be appropriate in a short student research project. The need for and procedure of assessing reliability and validity of research work and considering its generalisability are the subject of Chapter 14; this is an important chapter and should be given careful study. General advice on the conduct of research, including some guidelines on research writing and the structure of a dissertation, is given in the last two chapters. We also provide four appendices that may help you in the presentation of your dissertation and a fifth appendix on multiple-choice questions that give you a guide to your own understanding of the material in the book. You should also consult the world wide web for material on research methods and for specific examples of research work on topics which are of particular interest to the study of research. In addition, if time allows, you should visit the local university library to consult academic journals relevant to your programme of study, which will provide many ideas for research topics and a 'feel' for how research needs to be reported in an academic style. Even a couple of visits would be very worthwhile and time well spent.

We trust the reader will enjoy studying research methods, and that this book will provide you with the preparation, knowledge and skills, which will prove invaluable as you move along the pathway of research.

Good luck with your study of Research Methods!

John Adams
Hafiz T. A. Khan
Robert Raeside

Acknowledgements

This is the second edition of this book and the primary reason we have decided to produce a second edition is the very positive feedback we have received from students and teachers all around the world. We have taken on board many of these comments and added material, provided further explanation for existing material and included more examples and new datasets. We are very grateful to all the students and teachers who have suggested these ideas. We also wish to thank Dr Jesus Canduela for providing some very useful and important ideas for Chapter 10. We are also very grateful to Ms Nevine Essam, a student in her final year, for providing an excellent example of a research proposal, which is presented in Chapter 15. We are very grateful to Sachin Sharma at SAGE Publications for pushing us to finally produce this second edition of the book.

CHAPTER 1

Introduction to Research

1.1 INTRODUCTION

This book primarily aims to provide a clear discussion about the research methods employed in various disciplines related to our daily life problems. There is no absolute method that may be assigned in order to explore a particular research problem. Therefore, researchers may use different methodologies for investigating similar types of problems around the world. It does not, however, depend only on the cost and time involved, but also on the surrounding circumstances such as the availability of tools, mainly modern computer facilities, access to literature and publications and above all, dissemination of knowledge. The book has therefore been designed to illustrate research tools in a simple manner in a number of chapters, including formulating research, research design, data analysis and writing up the research results.

1.2 WHAT IS RESEARCH?

This appears to be a very simple question, but in fact it can depend on who is asked the question and it often depends on the subject of analysis. Research is a diligent search, studious inquiry or investigation or experimentation aimed at the discovery of new facts and findings; or broadly, it may relate to any subject of inquiry with regard to collection of information, interpretation of facts, revision of existing theories or laws in the light of new facts or practical ideas. More complex research would be required to investigate the causes of human fertility decline in Europe, or what could be the future labour force migration patterns in Europe.

Relatively simple research is merely aimed at acquiring the most basic type of information—but it is still research in a very real sense because it requires an individual to first identify the problem, then understand the problem, then know *where* to go for the information, then know *who* to ask for the information and also to know *what* questions to ask. If you think about it, failure to go through *any* of these basic steps will result in the research 'problem' remaining a research problem—and the individual concerned is very likely to miss his or her bus to work! Clearly not doing research properly has consequences!

A more academic approach to the question of 'what is research?' results in a more complex answer. Fundamentally, research is undertaken in order to enhance our knowledge of what we already know, to extend our knowledge about aspects of the world of which we know either very little or nothing at all and to enable us to better understand the world we live in. We can define a number of types of research studies that are aimed at achieving different knowledge outcomes:

- Descriptive research
- Explanatory research
- Predictive research

Descriptive research is aimed at simply describing phenomena and is not particularly concerned with understanding why behaviour is the way it is. This type of research is very useful for setting out baselines or 'templates' of how we think the world is. It is often the starting point of a research project into phenomena (known as an *exploratory* study) of which we know very little. For example, it aims to describe social systems, relationships between events, providing background information about the issue in question as well as stimulating explanations.

Explanatory research is deeper in the sense that it will describe phenomena and attempt to explain why behaviour is the way it is. In other words, it enables us to understand the very nature of what we are actually looking at. This type of research aims at explaining social relations or events, advancing knowledge about the structure, process and nature of social events, linking factors and elements of issues into general statements and building, testing or revising a theory.

Predictive research takes research one step further and is an attempt not only to explain behaviour but also to predict future behaviour given a change in any of the explanatory variables relevant to a particular phenomenon. If we can understand physical or human phenomenon then we will be in a much better position to predict its future path and possibly even to change it. This type of research is very important to governments in the design and application of policy.

In practice, most research work will include aspects of all three research 'types', although the third one is often the most difficult and problematic.

1.3 WHY IS RESEARCH CONDUCTED?

Research is conducted for a number of reasons, which in turn depend on the objectives of any particular 'research problem'. Of course, there are particular reasons for undertaking research at various levels to discover something new. As discussed earlier, it may be to find out something we do not already know or to enhance our understanding of phenomena that we already know something about. In the business arena, however, research tends to be undertaken in order to achieve one or more of the following objectives:

- To gain a competitive advantage
- To test new products and services
- To solve a management/organisational problem
- To provide information, which may help to avoid future business problems
- To forecast future sales
- To better understand shifts in consumer attitudes and tastes
- To enhance profitability
- To reduce operational costs
- To enable the management to prioritise strategic options for the future

One could go on and on with this list and we are sure that you can add to it. The main point, however, is that research (in whatever business or public sector organisation) is always undertaken for a clear purpose—to strengthen an organisation's ability to meet the demands of the future.

1.4 WHO DOES RESEARCH?

A very wide array of organisations and individuals do research. These range from the rather obvious such as market research companies through to the smallest government departments which need to know the impact of their work on the community. The following is just a small sample of the type of organisations/individuals who conduct research:

- Government departments
- Private companies
- Research companies
- Consultancy companies
- Academics

- Voluntary organisations
- Advertising agencies
- Market research companies
- And of course you, students!

The types of research each of the above undertake (descriptive, explanatory and predictive) will totally depend on the nature of the research 'problem' they are confronted with.

1.5 HOW IS RESEARCH CONDUCTED?

This is fundamentally related to the nature of the identified research 'problem'. For example, if the 'problem' is of a purely physical nature, it may be appropriate to undertake controlled laboratory experiments. This is the situation where the researcher can actually control the research environment to a significant degree. However, if the 'problem' is one relating to, for example, animal or human behaviour, it is much more difficult to control the research environment. In this case, it may be necessary (or even unavoidable) to conduct the research in a quasi-experimental fashion—that is, the researcher is able to control only a few aspects of the research environment such as the time of day to undertake observations or the sample from which to derive a generalised conclusion of the determinants of behaviour in a particular setting.

1.6 BUSINESS AND SOCIAL SCIENCE RESEARCH METHODS

There are several types of research and each type of research is associated with some sort of scientific tools and these will be discussed briefly later in this chapter. There is a common question to us: *Are business research methods different from others?* Business research deals with business phenomena such as price of commodity, supply of commodity, forecasting sales for a particular item, knowledge about market behaviour and marketing strategies to achieve a goal. And researchers can apply tools according to the essence of the inquiry. On the other hand, social and behavioural sciences deal with people who live in society, their culture and daily life. Social scientists thus follow a particular research strategy and apply the appropriate tools in order to fulfil the objectives of their study.

CHAPTER 2

Research Methodology

2.1 INTRODUCTION

This chapter will introduce the importance of research methodology in ensuring that research results can be generalised and if not then why not. It will then deal with approaches to social and business research and justify the importance of scientific approaches in research. A brief introduction to research ethics is provided at the end of the chapter.

2.2 RESEARCH METHOD VERSUS RESEARCH METHODOLOGY

The first thing to get absolutely clear about is that research method and research methodology are not the same thing! A research method is a way of conducting and implementing research. Research methodology is the science and philosophy behind all research. It goes into the heart of how we know what we know and allows us to understand the very strict constraints placed upon our concept of what knowledge actually is. Moreover, it allows us to understand the different ways in which knowledge can be created. This is especially important since if we know how knowledge and 'answers' to research questions can be created, then we are also in a position to understand what might be wrong with it. The concepts that underpin the subject of 'methodology' also enable us to be critical and analytical in the face of 'knowledge' being presented as 'fact'. Why should we accept the results of any research work at face value? We should not! The whole purpose of research is to extend and deepen our knowledge of the world, but if we are uncritical of how such knowledge was or is created, then we can never be in a position to improve its value to society.

2.3 APPROACHES TO BUSINESS AND SOCIAL RESEARCH

Researchers usually handle numerous problems and apply research methods to obtain the best guess answers to their questions. They may use a single study or a combination of two designs. The investigator has to decide about the types and combinations of research forms that best serve the goals of the study. Broadly speaking, there are two main domains of research frequently observed in the literature, which are Quantitative research and Qualitative research. The diverse practices and uses of today's research practices are listed in the following paragraphs.

Quantitative Research

This refers to the type of research that is based on the methodological principles of positivism and neo-positivism, and adheres to the standards of a strict research design developed prior to the actual research. It is applied for quantitative measurement and hence statistical analysis is used. Quantitative research is used in almost every sphere of life, such as in clinical, biological, epidemiological, sociological and business research.

Qualitative Research

This type of research uses a number of methodological approaches based on diverse theoretical principles (Phenomenology, Hermeneutics and Social Interactionism). It employs methods of data collection and analysis that are non-quantitative, aims towards the exploration of social relations and describes reality as experienced by the respondents. Qualitative research methods have long been used in the field of social sciences. For instance, these are the principal methods employed by anthropologists to study the customs and behaviours of people from other cultures, and are also used in such diverse areas as sociology, psychology, education, history and cultural studies. These methods have much to offer in studying the health and well-being of people and their daily lives in business and home.

Pure (Theoretical) Research

Pure research is usually used to develop new knowledge that advances our understanding of the real world. It evaluates concepts and theories and thus attempts to expand the limits of existing knowledge. It may also help in rejecting or supporting existing theories about the real world. In every sector of higher education there are some basic theories; a researcher's contribution in extending or improving any of these theories may be considered pure research (also known as theoretical research). Such research is very expensive and is usually carried out in government-funded projects by university research facilities or specific government laboratories. There is no obvious commercial value to the discoveries that result from pure research.

Applied Research

Applied research is conducted when a decision must be made about a specific real-life problem. The principal aim of scientists conducting applied research is to improve human conditions, although the results can have commercial value. It is directly related to social and policy issues. Examples of applied research include an investigation to improve agricultural crop production or a study on the development and commercialisation of technology with the potential to reduce carbon dioxide emissions. Types of applied research include action research (also known at times as evaluative research) and policy research.

Action research is a type of applied research. It is "the application of fact finding to practical problem solving in a social situation with a view to improving the quality of action within it, involving the collaboration and co-operation of researchers, practitioners and laymen" (Burns 1990: 252). It is actively involved in planning and introducing changes in policy, and researchers use their research expertise to monitor and possibly to evaluate its effects. It is also sometimes called evaluative research.

Policy research is ultimately concerned with the knowledge of action; its long-term aim is in line with the famous dictum that "it is more important to change the world than to understand it". This broad objective means that policy research encompasses a far more diverse variety of research, including theoretical research in many cases, but also descriptive research, which maps out the landscape of a topic, issue or problem, as well as reviews of how an existing policy is working. It can extend, in some cases, into formal evaluation research.

Longitudinal Studies

Longitudinal research involves the study of a sample (or cohort) on more than one occasion. In other words, longitudinal studies cover a long period of time, at times several decades, and follow the sample a repeated number of times. The longitudinal study is unique in its ability to answer questions about causes and consequences, and hence provides a basis for substantiated explanatory theory. It is commonly used in many disciplines. For example, in psychology, longitudinal studies are often used to study developmental trends across the life span; in public health, they are used to uncover predictors of certain diseases. Longitudinal studies include panel studies and cohort studies. A longitudinal study that involves collecting data from the same sample of individuals or households over time (usually regular intervals) is called a panel study. Panel studies take as their basis a nationally representative sample of the group of interest, which may be individuals, households, establishments, organisations or any other social unit. Longitudinal panel studies are conducted by educational organisations as well as by government institutions to study national income and expenditure. Panel members may be contacted by telephone, in a personal interview or by a mailed questionnaire.

Cohort studies sample a cohort in a selected time period and study them at intervals through time. A cohort is a group of individuals who experience the same event or share the same characteristics, namely, marriage cohort (individuals who got married during the same year or years), birth cohort (individuals born in the same year or years) and so on. In public health, such studies help us understand the causes of diseases and improve the overall health of individuals. Take for example, the UK-based Bradford birth cohort study that investigates why some children fall ill while others do not. It tracks the lives of more than 10,000 babies born in Bradford over three years from birth, through childhood, until they become adults.

Theory versus Empirical Study

Sound evidence is superior to argument based on false evidence, limited evidence or no evidence. Evidence has to be collected from the social world around us, and this requires that empirical research be conducted. 'Empirical', in this context, simply means 'based on evidence from the real world' in contrast to 'theoretical', which refers to ideas that are abstract or purely analytical. Theories must be tested against the real world. "Theory, in fact, is the building which is made from the hard-won bricks of the research studies" (Mann 1985). How can we collect sound evidence about the social world that can be used to increase our understanding of that world?

The purpose of science concerns the expansion of knowledge and the discovery of truth. Theory building is the means by which pure researchers expect to achieve the goal. It represents the real world and the events are supposed to be the reality. On the other hand, empirical study means the level of knowledge reflecting that which is verifiable by experience or observation.

A theory is a set of systematically interrelated concepts, definitions and propositions that are advanced to explain and predict phenomena (facts). A key element in our definition is the term 'proposition', which is linked with the term 'concept'. Concepts are an abstraction of the real world to allow us to more easily understand (by simplifying) the true nature of objects and events. Propositions are statements concerned with the relationships among concepts. A proposition explains the logical linkage among concepts by asserting a universal connection between them.

A hypothesis is a proposition that is empirically testable. It is an empirical statement concerned with the relationship among variables.

How Are Theories Generated?

Theory generation may occur at any level of the abstract conceptual level or at the empirical level. Theories may be developed with deductive reasoning by moving from a general statement to a specific assertion. Deductive reasoning is the logical process of deriving a conclusion from a known premise or something known to be true. At the empirical level, theory may be developed with inductive reasoning. Inductive reasoning is the logical process of establishing a general proposition on the basis of observation of particular facts.

2.4 JUSTIFYING THE SCIENTIFIC METHOD

The discussion that follows will help the reader in understanding some of these fundamental issues, which will always continue to surround all types of research. We proceed by considering the following:

- Styles of reasoning
- Common fallacies
- Useful quotations

Styles of Reasoning

In any type of research there are basically only two 'styles' of reasoning, that is, two methods of scientific enquiry: Inductivism and Deductivism. The inductivist method is strongly associated with the Scottish philosopher John Stuart Mill (1843) and the deductivist method developed in the early 20th century with Poincaré (1902) and later with Karl Popper (1934). The two styles of reasoning are presented below for comparison.

Induction

To draw general conclusions from a finite number of predictions implied by these laws J.S. Mill (1843). The method relies on empirical verification and was very popular in the 19th century.

Deduction

'Universal' laws are hypotheses to be 'tested' against the observations K. Popper (1934). The precise name is the hypothetico-deductive method—and still is used by many as the scientific method and largely replaced inductivism in the 20th century.

The two 'styles' are not opposite but tend to be complementary, and in conducting research people tend to swap between the two.

Inductivism

This approach to research relies on the empirical verification of a general conclusion derivable from a finite number of observations. That is, if an event repeats itself enough times then it can be concluded that the event will continue to occur ceteris paribus. This approach to research—observing the 'world' and coming to a generalisation about it—was popular in the 19th century and is still viewed by many as the scientific method. The inductivist method, therefore, operates from the specific to the general. Observation reveals patterns or trends in a specific variable of interest and these are then used to formulate a general theory of the nature and behaviour of that variable and often other variables fall into the same 'class' of phenomena. As Mill argued: "If we could determine what causes are correctly assigned to what effects, and what effects to what causes, we should be virtually acquainted with the whole course of nature." The narrow view of scientific enquiry and explanation forwarded by Mill became less popular in the 20th century with the rise of deductivism.

Deductivism

This has largely replaced the inductivist method and uses as its basis for the establishment of universal laws. These laws are essentially only hypotheses, which continue to require testing against the predictions of the laws themselves. That is, the universal laws remain so until one or more of their predictions are found to be false—in which case the theoretical framework, which derived them, needs to be revisited. The deductivist method thus operates from 'the general to the specific'. A general set of propositions relating to a given phenomenon is narrowed down to a specific set of testable hypotheses or to a single testable hypothesis. Testing the hypotheses requires the application of relevant data, which may or may not confirm the original arguments in the theory.

The research methodology purist would argue that the two 'styles' are so fundamentally different that either one or both of them are fundamentally flawed! However, in most research work it is necessary to use both, and in most research work they tend to be 'complementary'. However, each 'style' does contain dangers in the interpretation of research findings—they both are associated with common fallacies, which, unless given careful thought, tend to be assumed to be 'common' sense!

Syllogistic Reasoning (Deductive Logic)

A syllogism is a statement of two related parts from which a conclusion is drawn.
In general terms:

1. All objects A have the property B.
2. Object C belongs to the class of A.
3. Therefore, object C has the property B.

- Statements 1 and 2 are the premises of the syllogism.
- Statement 1 is the major premise.
- Statement 2 is the minor premise.
- Statement 3 is the conclusion.

The above is a Categorical Syllogism because its premises are Assertions. However, most syllogisms are hypothetical, that is, they take the form of If .... Then arguments. For example,

1. If A is true then B is true.   [If = antecedent]
2. A is true.   [Then = consequent]
3. Therefore, B is true.

For example, all crows are black—A is a crow; therefore, A is black.
The antecedent may have many clauses:

1. If A1 and A2, A3, ..., AN are true, then B is true.
2. If A1 and A2, A3, ..., AN are true.
3. Therefore, B is true.

The antecedent clauses are the Assumptions or tentative hypotheses and the consequent is the Prediction. Thus, the major premise is the Universal Law and the minor premise is the statement of relevant initial conditions.

The truth of a statement may be logically true in terms of the argument but will not necessarily be materially true, that is, the premises may be materially true or false and the conclusion logically true but a logically true conclusion must be materially true if its premises are materially true (Darnell and Evans 1990).

Fallacies in Deductive Reasoning

Two important (but common) fallacies are:

(a) *Affirming the Consequent*: This is a breach of the formal rules of logic. For example,

1. If A1 and A2, A3, ..., AN are true, then B is true.
2. B is true.
3. Therefore, A1 and A2, A3, ..., AN are true.

Here, the minor premise (2) is an affirmation of the consequent (not the antecedent). The rules of formal logic would require (3) to read as:
   3. Therefore, A1 and A2, A3, ..., AN are not necessarily not-true.
   For example, all crows are black; B is black. Therefore, B is a crow.
   The above fallacy is at the root of the problems involved in the verification of all theories.

(b) '*After this because of this*': If an event 'X' follows an event 'Y' then event 'Y' caused event 'X'. This is a seriously misguided assumption, which is commonly made, especially in the application of time-series analysis and regression models. It is tempting to presume that an action deliberately taken to achieve an outcome has 'caused' that outcome if the outcome actually occurs. For example, an advertising campaign designed to achieve a higher level of sales is often presumed successful if sales increase—however, what if sales did not increase? Was the campaign a failure? No. Because, in the absence of a campaign there is no way of knowing what sales would have done. That is, it is simply not possible to know, or logical to assume what would otherwise have happened. We cannot know the 'counterfactual' case, by definition!

Fallacies in Inductive Reasoning

Similar problems exist with inductivism, for example:

1. A recession has always followed a 'boom' in the business cycle.
2. What has happened in the past will continue to happen in the future.
3. Therefore, there will be a recession after the current 'boom' is over.

This syllogism shows the nature and fallacy of inductive reasoning. The essential problem with this methodological approach is the minor premise; it relies on the

Principle of Regularity, which in essence can only be a matter of faith! Regularity may be defined as the conformity to enforced rules and laws. The Scottish philosopher David Hume (1748) heavily criticised induction in his treatise *An Inquiry Concerning Human Understanding*. He argued that it is not possible to know that a 'regularity' of nature exists—but merely to consider that we should not expect all things to remain the same.

The material truth of the minor premise can never be demonstrated and therefore a universal statement cannot be logically derived (with certainty).

Falsification and Verification

Verification of theory is not possible but falsification is.

1. If A_1 and $A_2, A_3, ..., A_N$ are true, then B is true.
2. B is false.
3. Therefore, at least one of $A_1, ..., A_N$ is false and therefore the theory encapsulated by them is false.

This is a process of denying the consequent (3) since logically the antecedent (1) is false to begin with.

This of course means that in arriving at a conclusion which supports a hypothesis we cannot be bullish in our claims of what we have 'discovered'. The limits of validity of our findings need to be clearly spelled out, particularly where we are tempted to generalise beyond the 'class' of problem under investigation.

Common Fallacies

These are simple errors that are easy to make unless one thinks more carefully about the logic of a statement and how it has been derived.

For example, one might make the statement that:

1. All unemployment is voluntary.
2. A. N. Other is a member of the unemployed.
3. Therefore, A. N. Other is voluntarily unemployed.

Statements 1 and 2 are the major and minor premises, respectively, of this argument. Statement 3 is the conclusion. However, note that statement 3 is merely an assertion based on the assumption that the major premise is a fact and not a premise! This in fact breaks the formal rules of logic.

This type of fallacy is often found in deductive reasoning. Given that this is indeed a very common fallacy, then you should be aware that, in interpreting your own research findings, you are likely to make the same error unless you think very carefully about the research question(s) and how you arrived at the answers. This also applies to your critical review of a published research paper—carefully consider if the conclusions make logical sense in terms of the questions or hypotheses set out at the beginning. This is not always an easy thing to do but it is very important to try!

A different type of fallacy can be found in inductivism where the bedrock of empiricism was laid. Here is an example:

1. Financial innovations (new products) have always found/created new markets.
2. This will always be the case.
3. Therefore, more financial innovation will find/create more new markets.

The problem (or fallacy) here is that the whole argument rests on a belief that the past is a guide to the future; that there is some principle of regularity or determinism at work. Hence, it is the minor premise (statement 2) which is fundamentally flawed in logic.

In designing a research project and implementing it, you should bear in mind that it is easy to fall into the trap of making statements or arguments which may appear reasonable but in fact are logically incorrect. This is particularly the case when interpreting data, whether quantitative or qualitative. Consider the following paragrapgh.

The premises of an argument may be materially true or false and the conclusion derived logically true but a logically true conclusion cannot be materially true if its premises are materially false. If the premises of the argument are materially true then the conclusion derived must be materially true. That is, the truth of an argument may be logically true in terms of the argument, as it has been set out, but will not necessarily be materially true (see Darnell and Evans 1990). So, what does this mean in terms of undertaking a research project for your degree?

*It is not possible to verify a theory, only to falsify it or elements of it.
That is, be modest in your claims of what you have found!*

*Do any of your findings make sense in terms of being logically true
(from your assumptions) but they cannot possibly be materially true?*

*A methodology must be logically sound and not simply a set of
methods of information gathering.*

All of the above lie at the root of the Rules of Evidence whether applied in mathematics, social science, business or a courtroom! There are four of these, discussed in the following sections, but each contains 'sub-rules' that may apply to each other type.

Real Evidence

Real evidence is a thing wherein the existence or characteristics of which are relevant and material. To be acceptable real evidence must be relevant, material and competent. These three criteria are applied in order to minimise the tendency to break formal rules of logic in relation to a 'thing' being materially true. In this regard *qualitative data* is very weak (i.e., often non-replicable).

Demonstrative Evidence

This involves offering an illustration to support an argument (e.g., a diagram, a quotation); however, it is the evidence that can easily be shown to be materially false even if its premise is logically true.

Documentary Evidence

It is often seen as a type of real evidence but will normally require authentication by other evidence. For example, a completed questionnaire is a form of documentary evidence but what it contains is not necessarily materially true. A death certificate is not in itself evidence of death—but merely an authentication of real evidence that a person is actually dead. A current example of the weakness of such evidence is passports—just because I have a passport with a name and a photograph of me does not materially prove that it is actually me!

Testimonial Evidence

This is the most basic type of evidence and often does not require authentication per se. However, the existence of bias, interest, prejudice and other grounds that would raise doubts about the credibility of a respondent can only contribute to the weight of their 'response' and do not affect its competence. However, this type of evidence, more than the others, suffers from a greater probability that its content is not in fact materially true. Basically, there are two golden rules:

1. It is not possible to verify a theory, only to falsify it or elements of it. That is, be modest in your claims of what you have found!
2. Do any of your findings make sense in terms of being logically true (from your assumptions) but they cannot possibly be materially true?

The latter 'rule' is particularly relevant to a common fallacy found in the literature on the role of small businesses in job creation—it is sometimes referred to as the Size Distribution Fallacy. Consider a typical definition of a small business (to be found in many countries) as being a firm that employs less than 50 people and consider the data in Table 2.1.

Looking at column 4, it would appear that small firms had created nine jobs between Year 1 and Year 2 while large firms only managed the creation of one more job in the same period. In fact this conclusion is materially false even although, given our definition of a small firm, it is logically true! The latter is the case simply because one large firm has been 'redefined' as a small firm due to it falling through the arbitrary and 'magic' number of 50 employees. Data presented like those above easily lead to the common fallacy that small firms create more jobs than large firms—but it is materially false in the above example (and more often than not in the 'real' world).

So what does the above data actually tell us? Simply that one of the large firms lost 21 jobs, and so by definition became a small firm, the largest firm created 56 jobs and the small firm lost 25 jobs. So, in terms of before and after, the two large firms created a net 35 jobs, the small firm lost 25 jobs and the local economy gained a net 10 jobs overall. What the above example shows is that the definitional basis of many 'everyday' statistics relevant to finance, economics, business, management, marketing and population trends and many other phenomena are often the source of the problems encountered in interpreting the real world.

Useful Quotations

Here are three apt quotations which should be borne in mind when arriving at any conclusions in research. Together, they represent a useful 'check' on any conclusion you arrive at, given the original premises, assumptions, theory base and empirical data upon which your research has been constructed.

1. All models are wrong, but some are useful—Box (1979) (Statistician)
    That is, we do not and cannot 'know' all there is to know even from a very narrowly focused piece of research.

Table 2.1 Job Creation and Small Firms

	Firm 1	Firm 2	Firm 3	Small Firms	Large Firms	Total
Year 1	30	55	65	30	120	150
Year 2	5	34	121	39	121	160
Net difference	−25	−21	56	9	1	10

SOURCE: Davies, Haltiwanger and Schuh (1993).

2. There is no logic of proof, but there is a logic of disproof—Blaug (1993) (Economist)

   Because of (1) something is bound to turn up and shatter a 'proof'—hence it is more fruitful and more logical to attempt disproof.
3. An approximate answer to the right question is worth a great deal more than a precise answer to the wrong question—Tukey (1962) (Mathematician)

Point 3 is a reaffirmation of 1 and a recognition that 2 is logical, since searching for disproof only makes sense if we know our 'answers', by definition, cannot be precise. It is useful at this point to summarise the key arguments discussed above. This is done using a set of definitions, which you should consider in the process of your study in your preparation and implementation of any research dissertation whether at undergraduate, Masters or PhD level.

Methodology: '... is enquiry into why the accepted is judged acceptable' (De Marchi and Gilbert, *Oxford Economic Paper*, 41, 1989).

Science: 'A field of enquiry divorced from its subject matter' (Anon, 19th century).

Explanation: a subjectively determined activity.

Models: are a representation of 'reality' and by definition incomplete and tentative.

Theory: is a 'formalised set of concepts, that organise observations and inferences and predicts and explains phenomena' (Graziano and Raulin 1996). To be scientific, a theory must be testable. A good theory requires a strong empirical base. Theories are developed by using inductive and deductive logic.

Constraints: include limited knowledge, access, ability and time.

If you have a good understanding of the above, then you will be in a position to undertake a good critical review, prepare a workable and practicable research plan for your dissertation and be in a much stronger position to be able to make reasonable statements from and to identify the limits of validity of your own research.

Let us return to the concept of the 'scientific' method. Across the social sciences and within intellectual and philosophical debates there remains a great deal of disagreement over what constitutes 'science' in the sense of what constitutes actual and accepted knowledge. Possibly the one single approach to the discovery of knowledge, Positivism, has been more criticised than any other. However, the criticisms tend to mainly come from other 'approaches', which themselves are highly dubious. Not only are these other approaches dubious in terms of the discussion above (syllogistic reasoning, material truth, rules of evidence) but they also can be seen as nothing more than fashionable 'fads' which come and go with the times. Four of these in particular can be demonstrated to have followed a typical 'fad' curve although there are others.

An academic 'fad' is simply a trend which reflects academic interest (and publication) in a range of subjects but investigated in a manner that was previously not applied or applied by very few. So it is an 'idea' that catches on but later declines as its ability to explain the world becomes increasingly doubted by other academics. The evidence for this argument (for just four methodological fads) is given in the following figure:

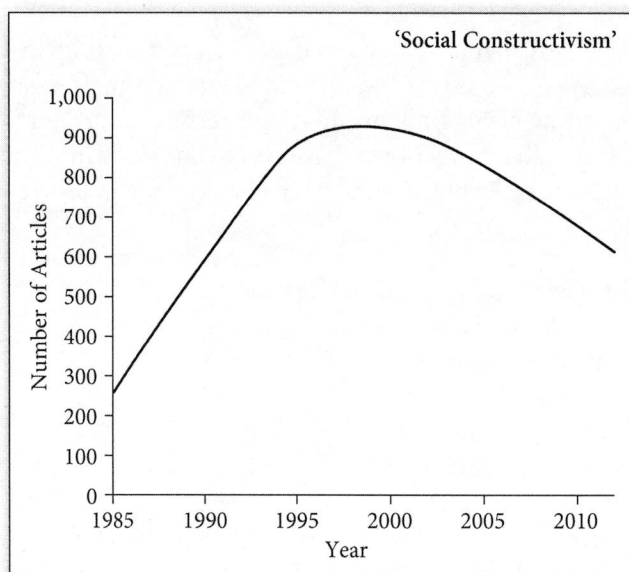

'Social Constructivism'

This 'scientific method' purports that knowledge comes from the 'social construction of reality'—that is, nothing is an actual given, the world is constructed by our relations, perceptions and beliefs. This was very popular with sociologists until the middle 1990s. Since then the term has been used less and less in academic journals and is now in steep decline. The reason is simply because it has been rejected as spurious in logic and baseless in its concept of what constitutes a refutable scientific theory.

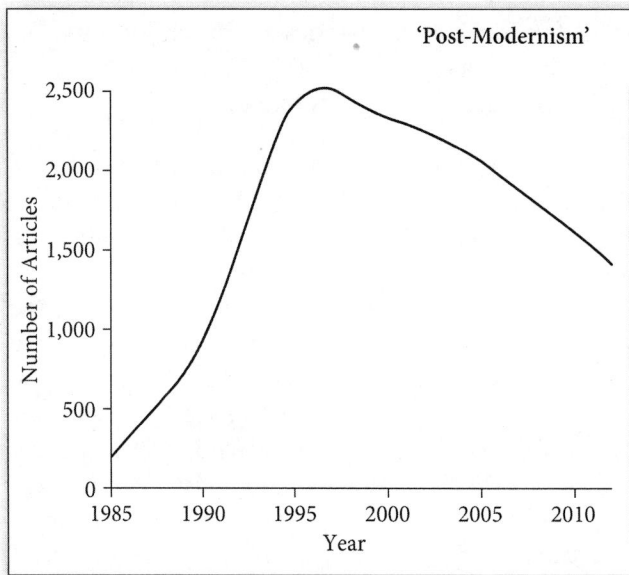

'Post-Modernism'

Post-modernism, just like social constructivism, is a group of ideas that argue that Positivism (the world is the way we find it) is unsustainable. Therefore, the 'post-modernists' argue it needs to be changed (this is also the Marxist approach). But changed to what? In fact the roots of this fad can be found in the literature that gave Adolf Hitler many of his Nazi ideas in the 1920s! This 'methodology' has no redeeming features in logic, evidence or anything else.

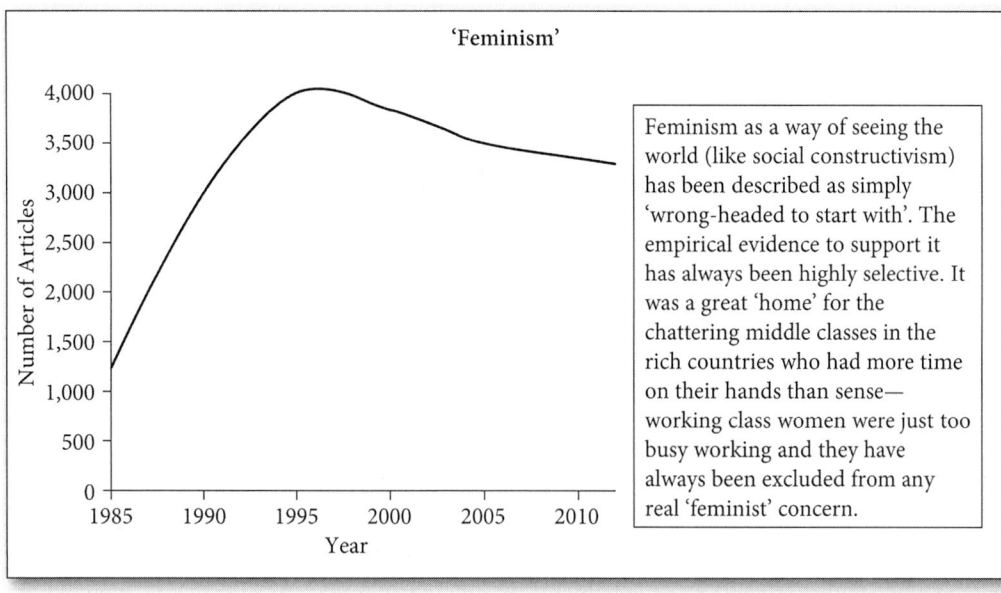

'Feminism'

Feminism as a way of seeing the world (like social constructivism) has been described as simply 'wrong-headed to start with'. The empirical evidence to support it has always been highly selective. It was a great 'home' for the chattering middle classes in the rich countries who had more time on their hands than sense—working class women were just too busy working and they have always been excluded from any real 'feminist' concern.

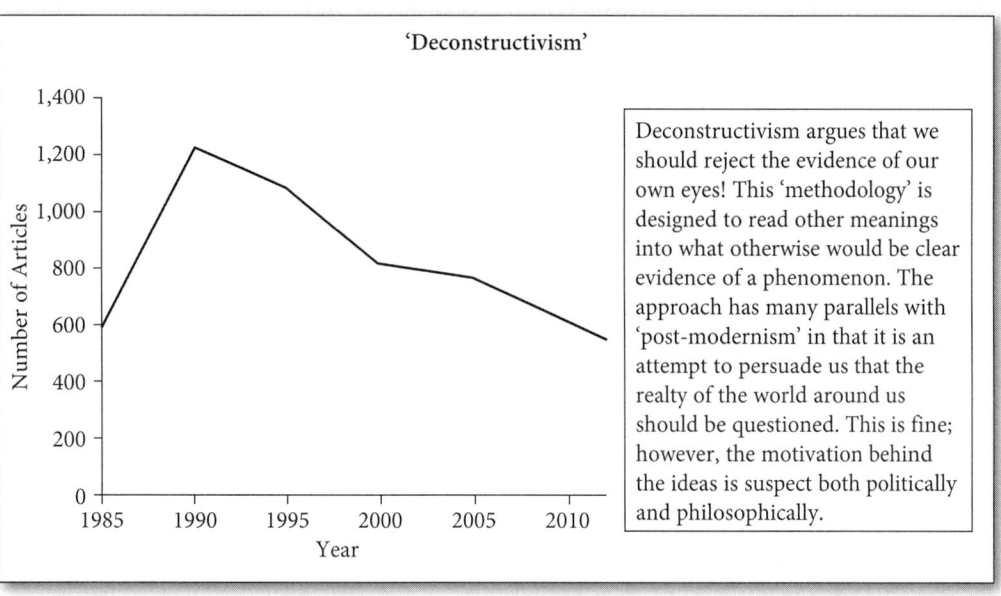

'Deconstructivism'

Deconstructivism argues that we should reject the evidence of our own eyes! This 'methodology' is designed to read other meanings into what otherwise would be clear evidence of a phenomenon. The approach has many parallels with 'post-modernism' in that it is an attempt to persuade us that the realty of the world around us should be questioned. This is fine; however, the motivation behind the ideas is suspect both politically and philosophically.

SOURCE: JSTOR Article counts from 1985 to 2012.

There are many other 'isms' that academics have convinced themselves need to be 'taught' to students. In fact we would argue that teaching this 'stuff' is a terrible waste of a student's time, often a reflection of our own ignorance of what these 'ideas' really are and a complete distraction from what research and dissertation students should be doing—concentrating on the world's problems as they find them by applying sound, logical, scientific enquiry. In other words, it applies the Scientific Method (Positivist) approach through qualitative or quantitative research methods or a combination of both. So, what is *Positivism*? Is it just another 'ism'? The answer to this question is a demonstrable *No*. The following quotation provides one of the clearest definitions of what the scientific methodology of positivism actually constitutes:

> *Any sound scientific theory, whether of time or of any other concept, should in my opinion be based on the most workable philosophy of science: the positivist approach put forward by Friedrich Hayek and Karl Popper and others. A good theory will describe a large range of phenomena on the basis of a few simple postulates and will make definite predictions that can be tested. ... If one takes the positivist position, as I do, one cannot say what time actually is. All one can do is describe what has been found to be a very good mathematical model for time and say what predictions it makes.*
>
> Stephen Hawking (2001: 31)

However, can positivism be applied in the Social Sciences? Most scientists (and economists and psychologists) would argue it is simply the only *scientific method* we really have. They would argue two points and other 'isms' would counter the arguments as follows:

1. The world exists independently of people—our job is to understand it and explain it.
   Criticism: social and personal interaction with the 'world' affects our research outcomes; therefore, the world does not exist independently of people.
   Not true: in physics, biology, computer systems engineering and economics and psychology we have always understood that our 'interaction' with the world has a profound effect on research outcomes. Well-known examples are the Heisenberg Uncertainty Principle, policy bias in surveys and the completion of tax returns!
2. Positivism creates 'laws' of regularity that can be used to predict natural and human behaviour.
   Criticism: Positivism claims these are universal and generalisable over time, which cannot be materially proven.
   Not true: No serious academic understands that our theoretical knowledge is wholly limited and that theories need to be constantly refuted, revised or created.
   These 'criticisms' of positivism (the true scientific method) reveal a genuine ignorance of what the scientific method actually does. This also raises another important question in any type of research—how can we be certain that our research approach is ethical in the sense that we actually use all the evidence

available and do not 'select' just the evidence that supports our personal or political view of the way we would prefer the world to be?

2.5 RESEARCH ETHICS

In conducting any research there is an ethical responsibility to do the work honestly and with integrity. If you do not conduct your work in an ethical manner, you will fail. This will apply to all stages of the research cycle. If the work is not honestly undertaken then it will be essentially fraudulent and fraud must be avoided in research—it can come in several forms:

- Being selective in sampling.
- Not reporting survey response/participation rates.
- Deliberately biasing the data collection instruments—for example, asking leading questions in surveys.
- Making up data—because you cannot be bothered doing the data collection.
- Falsifying results—to make them fit your conclusion.
- Trimming—removing data that does not fit in with your analysis. This may be a legitimate thing to do but you must make it clear what has been done and why.
- Biased or inappropriate analysis.

The choice of techniques for carrying out analysis can lead to widely varying conclusions. This is often the case when the data is skewed, as for example, in salary data. Reporting the mean salary would not be as ethical as reporting the median. Consider the following example of survey data. A question was asked, 'Do you think nuclear power is a good thing?' This was rated on a 1–9 scale, with 1 being very bad and 9 very good. The data obtained is as follows:

Very Bad	1:	1
	2:	1 1 1
	3:	1 1 1 1 1 1 1 1
	4:	1 1 1 1
Neutral	5:	1
	6:	1 1 1 1
	7:	1 1 1 1 1 1 1 1 1
	8:	1 1 1
Very Good	9:	1

Clearly to use the mean of around 5 would not legitimately portray the true story.

Obfuscation

It is obscuring the research findings by the reporting style—for example, by not highlighting the results that are important or those that the researcher does not like. This would be done by not reporting information that is contradictory to your conclusions, or by hiding information by producing a very verbose report full of very technical/specialised terminology or long paragraphs with long-winded sentences. Obfuscation also includes the display of data. Graphs must be given titles, labelled axis and any transformations used, such as logging the axis, must be clear to the reader.

Plagiarism

This is passing off someone else's work as your own. This is unacceptable and any quotes and illustrations used must be attributed to their source and properly referenced.

To avoid many of these problems, research must be carefully planned and one should strive to avoid cutting corners. Research ethics is summed up in the following statement: "It is unethical to conduct research that is badly planned or poorly executed" (Declaration of Helsinki 1975).

In planning research, the ethical consequences to the individual and to society must be considered and made clear. If you are involved with humans as subjects in experiments or as cases in a survey, informed consent must be obtained. Therefore, you need to tell the subjects about your research, what you hope to achieve, how they will be affected, and to ensure that they understand. There is also, where relevant, an obligation to ensure privacy and confidentiality. This can relate to the information collected, individuals involved, the setting and how the research data and findings are stored and disseminated. There are some ways of ensuring this, such as by putting passwords on files, limiting copies, aggregating findings, removing scales, labelling people/companies as a, b, c, etc., and making small changes at random to data in tables to prevent individuals from being identified—a process called Barnardisation.

General principles for good practice are:

- *Design your research well*: Do not use vulnerable groups such as the young (16 years or less), the old (over 70 years) and those with disabilities. Ensure that you obtain informed consent. To do this your research aims must be clearly explained and understood by participants—you should not deceive your subjects. You need to ensure that the subjects have understood. Often a consent form is produced, which research subjects would be asked to sign. In such letters the purpose and design of the research should be explained clearly and how data will be stored and used should also be explained. A contact should be provided if the subject requests for more information. No one should be coerced to take

part. Coercion can be subtle, such as offering incentives or bribes to participants—avoid giving incentives.
- If confidentiality is required then the research has an obligation to ensure that subjects cannot be identified. Generally the anonymity and privacy of subjects should be protected.
- Protect the data—do not leave it lying around; keep data in a secure place and carefully enter it into computer files if required. Always keep back-ups—but keep these securely. Have a data-disposing policy—when the data is no longer needed, it should be destroyed.
- Ensure that no harm can come to participants or to the researcher (including anyone who may be helping you). Do not falsify results or make exaggerated claims. Report on reliability and validity—see Chapter 14.
- Many professional bodies have produced codes of conduct, which, if followed, are designed to avoid many of the ethical pitfalls.

  Good examples are the British Sociological Association, see http://www.britsoc.co.uk/about/equality/statement-of-ethical-practice.aspx, and the code of conduct of the Royal Statistical Society's Code of Conduct, which can be found at http://www.rss.org.uk/codeofconduct.

  For further information, Frankfort-Nachmias and Nachmias (1996) provide a good treatment of ethics in social research and the Economic Social Science Research Council (ESSRC) provides a robust Research Ethics Framework—this should be consulted. Another useful source is the listing of UK codes of conduct, found in the *Research Ethics Guidebook* located at: http://www.ethicsguidebook.ac.uk/Professional-Guidelines-313.

2.6 EXERCISES

Exercise 1

Which of the research approaches outlined in this chapter would be suitable for the following research topics?

1. Measuring the success of stock market forecasts
2. Identifying factors affecting the development of small businesses
3. Measuring the suitability of partners to outsource
4. Identifying the best method of recruitment
5. Designing and evaluating training programmes
6. Measuring the motivation and happiness of employees

Give reasons for your answers and discuss any practical problems you may encounter with your chosen research approach.

Exercise 2

Discuss how you might conduct field research if researching the play patterns of children at home and in school. What particular problems would you encounter in gathering suitable data? What ethical issues will need to be addressed?

Exercise 3

Read over the ethical codes of conduct links included in this chapter and discuss the similarities and any differences between them.

Which issues are likely to be particularly relevant to your own future research, with respect to both completing your programme of studies and your future career?

Exercise 4

Review a suitable journal article (related to your own area of expertise) in terms of the ethical issues addressed by the researcher. What other ethical issues might the researcher consider?

2.7 REFERENCES

M. Blaug, *The Methodology of Economics—Or How Economists Explain* (Second edition) (Cambridge: Cambridge University Press, 1993).
G.E.P. Box, 'Robustness in the Strategy of Scientific Model Building', in *Robustness in Statistics*, ed. R.L. Launer and G.N. Wilkinson (New York: Academic Press, 1979).
R.B. Burns, *Introduction to Research in Education* (Melbourne: Longman Cheshire, 1990).
A.C. Darnell and J.L. Evans, *The Limits of Econometrics* (Aldershot: Edward Elgar, 1990).
S. Davies, H. Haltiwanger and S. Schuh. 'Small Business and Job Creation: Dissecting the Myths and Re-Assessing the Facts'. NBER Working Paper, No. 4492, October 1993.
Declaration of Helsinki, Recommendations Guiding Medical Doctors in Biomedical Research Involving Human Subjects, http://ethics.iit.edu/codes/coe/world.med.assoc.helsinki.1975.html, 1975.
N. De Marchi and C. Gilbert, *History and Methodology of Econometrics* (Oxford: Clarendon Press, 1989).
C. Frankfort-Nachmias and D. Nachmias, *Research Methods in the Social Sciences* (Fifth edition) (New York: St. Martin's Press, 1996).
A.M. Graziano and M. Raulin, *Research Methods: A Process of Enquiry* (New York: Longman, 1996).

S. Hawking, *The Universe in a Nutshell* (London: Bantam Press, 2001).
D. Hume, 'An Enquiry Concerning Human Understanding', in *Enquiries 1975,* ed. L.A. Selby-Bigge and P.H. Nidditch (Oxford, 1748), http://www.infidels.org/library/historical/david_hume/human_understanding.html
P.H. Mann, *Methods of Social Investigation* (Second edition) (Oxford: Blackwell, 1985).
J.S. Mill, A System of Logic: Ratiocinative and Inductive (1843), http://cepa.newschool.edu/het/profiles/mill.htm
H. Poincaré, *Science and Hypothesis* (Reprint 1952) (New York: Dover, 1902).
K. Popper, *The Logic of Scientific Discovery* (Second edition 1968) (London: Hutchinson, 1934).
J.W. Tukey, 'The Future of Aata Analysis', *Annals of Mathematical Statistics* 33 (1962): 1–67.

CHAPTER 3

The Research Cycle

3.1 INTRODUCTION

In this chapter we walk through a process of organising, planning, conducting, analysing and reporting on research called the scientific method. This is a suitable approach for research at any level of study (Masters or PhD), as it is very goal-oriented and focused. It helps you get your research completed in a timely manner. However, it has been criticised for being reductionist and preventing the researcher from understanding the whole nature of the research problem. This process is described here as a 'research cycle', a term advanced by Karl R. Popper in 1979.

3.2 THE RESEARCH CYCLE

The research cycle is the application of the scientific method as displayed in Figure 3.1. This is drawn from Popper's (1979) original exposition of this systematic procedure. The steps in the cycle are detailed in the following sections.

Formulate

The key stage is the initial understanding of the problem or situation to be modelled. This is the formulation phase. In reality, for a model to be any good, a great deal of time and effort must be devoted to this phase; otherwise, the model will be analogous to a house built on sand. In this phase, the 'stakeholders' in the research problem are identified, in order to place boundaries on the problem. The variables and emotive issues are

Figure 3.1 The Research Cycle

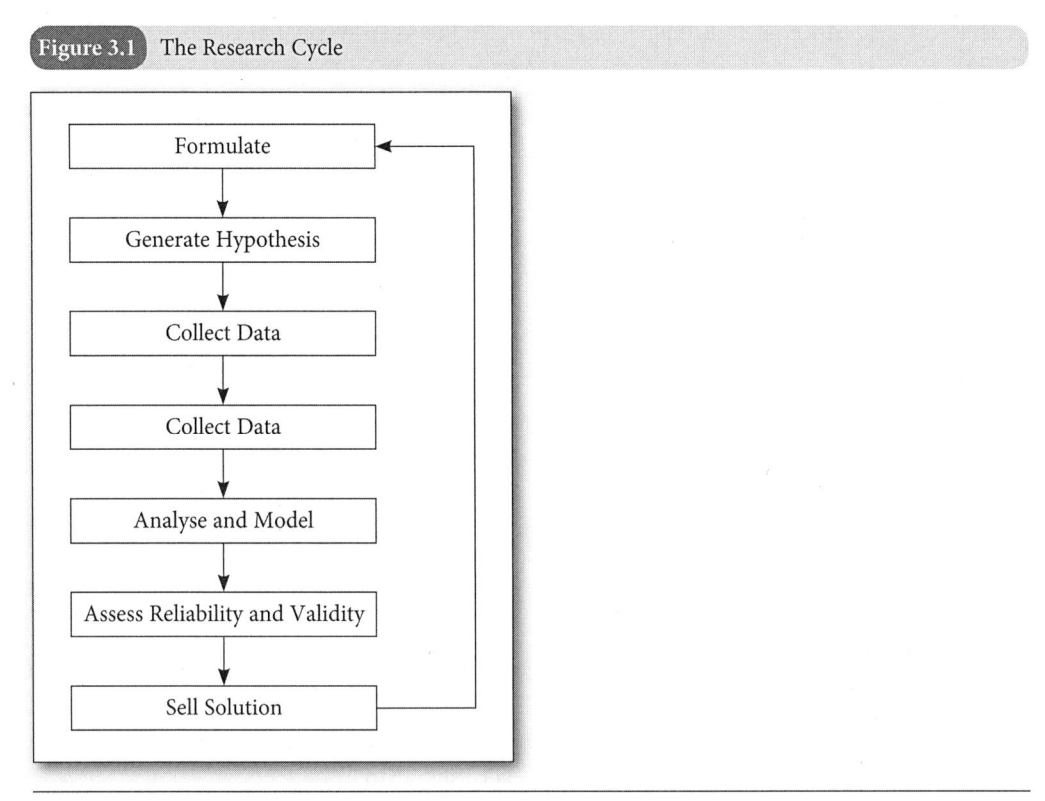

```
Formulate
   ↓
Generate Hypothesis
   ↓
Collect Data
   ↓
Collect Data
   ↓
Analyse and Model
   ↓
Assess Reliability and Validity
   ↓
Sell Solution
```
(Sell Solution loops back to Formulate)

SOURCE: Authors' own.

also identified; this occurs through protracted discussions and brainstorming activities. The literature is extensively reviewed to identify how others have undertaken similar research, what variables they have used and how measurements are made. Hypotheses or propositions are formed as to how the variables might influence one another. For example, if we were attempting to predict the monthly sales of a particular product in a national market we might consider the following model:

$$\text{Monthly sales of Product } X = a + b_1 \text{ Price } (X) + b_2 \text{ Price } (Y) + b_3 \text{ Income} + b_4 \text{ Advertising}$$

where 'a' is called the constant of the model and represents the sales level if none of the other variables are present. The 'b's are the coefficients of the model, which represent the contributions of the selected independent variables. The variables Price (X) and Price (Y) can be measured on a continuous scale and are hence called *continuous* variables. The variables Income and Advertising are sometimes used as categorical variables and are used to represent different income groups and different levels of advertising expenditure on X, such as high and low. Some ideas on successful problem-solving can be found in the soft systems methodology literature; see, for example, Checkland and Scholes (1990). In particular, it clearly sketches out key questions of the research problems.

Generate a Hypothesis

Conventional hypothesis statements help guide the analysis. They take the form of the null hypothesis (H_n), that there is no effect or relationship, which is compared to the alternative hypothesis (H_a), that there *is* an effect or a relationship. To consider the previous example, one would write:

> H_n: Sales levels per month of X are not significantly different between high levels of advertising expenditure and low levels.
> versus
> H_a: Sales levels of X per month are significantly higher when advertising expenditure is high.

Collect Data

If one can assume for the moment that the speculative model is acceptable, then data collected on selected variables are usually classified into *independent* and *dependent* types. Here, for example, the independent variables are the monthly price of X, monthly price of Y, income groups and advertising expenditure. For the same period, the actual sales of product X are recorded, which is the dependent or *response* variable. An independent variable is a variable that is expected to influence the dependent variable.

Explore the Data

This data is then explored, often using very simple tools, such as scatter plots of sales against each of the independent variables. Another reason for this exploration of the data is to catch any oddities that exist in the data, such as abnormally high sales (e.g., due to Christmas) or data-entry errors.

Analyse and Model

Assuming that relationships exist and there are no problems in the data, the data is then analysed to explicitly answer the research questions and accept or otherwise reject the hypothesis. Statistical methods might be used such as analysis of variance or regression or the analysis might be qualitative in nature. In either case, a model can be formed on how relationships have been found to work and this would allow 'what if' investigations to be made. Again, the models can be statistical such as those arising out of the use of regression or they might be schematic, drawing attention to the nature of how themes might affect an outcome.

Assess Reliability and Validity

How reliable the data is in terms of how well it was collected, measured and free from bias and corruption should be assessed. What was the degree of subject–researcher interaction? Was the subject being led or were they trying to please the researcher? Assessment of validity hinges around the degree to which the researchers are measuring what they set out to measure—it relates back to the research design and the theoretical underpinning of the measures and how well the measurement was conducted. A question often asked is "what alternative explanations are there for the findings?" How well the research stands up to this scrutiny is a measure of the quality of the research and this is a vital stage.

Sell Solution

Next comes the 'sell the solution' phase. This is an addition to Popper's formulation. It is incorporated here to avoid what has been a major failing of researchers in the past. This was a failure of communication, and great care and time must be taken to explain to users how the model has been arrived at and how it works. Similarly, those who are affected by the consequences of decisions that arise out of the use of the model must be able to understand how and why the model is appropriate. This phase must not be underestimated. Reference should be made to the previous step and the assessment of the quality of the research. The more reliable and valid the research can be shown to be; the more useful it is, the more people will accept it. If at any stage in the research cycle there are unsatisfactory findings, one returns to the formulation phase.

In summary, the main research stages are:

1. Specify the real *problem*.
   - Investigation purpose (hypotheses to be tested)

2. Set up a *model* (from *theory*).
   - Break the problem (1) into parts

- Select variables
- Make *sensible* assumptions
- Determine the *limits* of *validity*

3. Formalise the model.
   - Mathematical or statistical version of the problem (if appropriate)
   - Find the relationships involved
   - Find a functional form

4. *Solve* the problem.
   - Choose appropriate techniques
   - Amend the model and/or change the approach if needed

5. Interpret the results.
   - Does step 4 make sense in terms of step 1?
   - If not, *why not*?

6. *Validate* the model.
   - How realistic were the initial assumptions?
   - What if one or more were changed?
   - More refining needed? At what *cost*?
   - If *yes* (and it is affordable), then back to step 2!

7. Generalisation: can the model and results be generally applied to this *class* of problem?
   - If *yes*, then a contribution to *knowledge* has been made.
   - If *no*, then a *major* problem has occurred somewhere along the line!

8. Report the findings.

3.3 PROBLEMS WITH THE RESEARCH PROCESS

The above is essentially a reductionist approach and one should be aware that there are many inherent problems in trying to depict reality using such a schematic procedure. The research process can be thought of as a series of screens through which progressively less of the data passes; much data passes through the collecting screen (the measuring tool, e.g., a questionnaire) or is reflected away (conscious decisions not to take part) and what does pass through can be distorted, and that which reaches the final user may be of little use. This is illustrated in Figure 3.2.

Figure 3.2 Reduction of the Issues

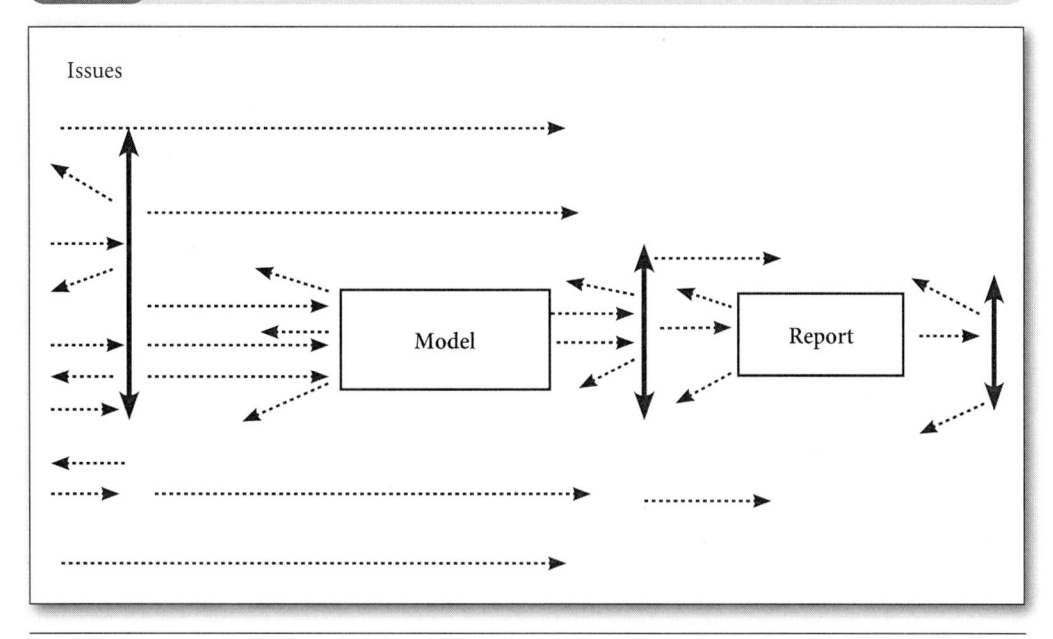

[Diagram: Issues → Model → Report, with arrows showing data being reflected, refracted, and passing through screens at each stage]

SOURCE: Authors' own.

The first screen represents the data collection phase, where valuable data is often missed, some being reflected by the measurement system, as would be the case with a poorly designed questionnaire. On passing through the screen, the data can be distorted (refracted) due to measurement and transcription errors. For example, suppose the data input to a computer was to be '12345' but the operator actually input '12435': a transcription error has occurred. This is a very easy mistake to make, which, if not detected, is capable of undermining any data analysis you may undertake. This error can be repeated at the modelling stage and lead to misinterpretation at the report stage and, finally, those who read the report absorb even less of the original message. And if the types of errors discussed have been made, the reader might even get an entirely wrong message!

There are also elements of choice within research that are often subjective and can be at the whim of the researcher.

Here, decision-makers choose the variables to enter into the model. The list of potential variables is often more a matter of the imagination than anything else. Once the variables are listed, they are often selected subjectively (see White 1975).

The points raised in this chapter are really the subject of the forthcoming topics where you will be introduced to various research tools. This will allow you to conduct good research on your own, and critically appraise research done by others. Before moving on, try the exercises below to check your understanding of this chapter.

3.4 EXERCISES

Exercise 1

Which of the research approaches outlined in this chapter would be suitable for the following research topics?

1. Measuring the success of stock market forecasts
2. Identifying factors affecting the development of small businesses
3. Measuring the suitability of partners for outsourcing
4. Identifying the best method of recruitment
5. Designing and evaluating training programmes
6. Measuring the motivation and happiness of employees

Give reasons for your answers and discuss any practical problems you may encounter with your chosen research approach.

Exercise 2

The model formulated to describe the relationship between monthly sales of product X and prices, incomes and advertising expenditure seems perfectly reasonable.

Give reasons why you would not expect this model to be a particularly good predictor of monthly sales of product X.

Exercise 3

What problems might you encounter in

1. Providing a definition for the above variables
2. Gathering appropriate data on the above variables

3.5 REFERENCES

P. Checkland and J. Scholes, *Soft Systems Methodology in Action* (Chichester: Wiley, 1990).
K.R. Popper, *Objective Knowledge: An Evolutionary Approach* (Revised edition) (Oxford: Oxford University Press, 1979).
D.J. White, *Decision Methodology: A Formalization of the OR Process* (Chichester: Wiley, 1975).

CHAPTER 4

Literature Review and Critical Reading

4.1 INTRODUCTION

In this chapter we consider the importance of undertaking a literature review, what the review should do for you, how it links to your research questions, your research method(s), your theoretical framework and your findings. By the end of the chapter you will also appreciate why it is always the case that a literature review is never complete. This chapter provides detailed explanations of how to undertake a literature review and why it is a pivotal element of any research enquiry. You will be reminded of some of the discussions in Chapter 1 and given pointers to how the literature review relates to later chapters in the book. At the end of the chapter, you are presented with an example of a literature review section from a published paper, which exemplifies the key elements required in a literature review. The present chapter is structured as follows:

- The importance of a literature review
- What should the literature review do?
- Types of literature review
- Some general points in literature reviewing
- Obtaining literature sources
- Searching the literature
- Assessing the quality of literature
- An example of a literature review
- Critical reading

The inclusion of a section especially on critical reading will improve the overall undertaking of the literature review. However, given the importance of the skills required for undertaking a literature review and for engaging in critical reading, it is necessary that they are given sufficient space on their own. They are of course complementary to each other. This will particularly help in preparing your first assignment for the course—an assignment on critical reading of an academic piece of research published in an academic journal and selected by you.

4.2 THE IMPORTANCE OF A LITERATURE REVIEW

In any research project it is essential to understand what has already been done (if anything) in the specific topic you have chosen and what has been done in the wider subject area of that topic. This is essential for several reasons and the importance of a literature review can only be appreciated when we ask ourselves a number of specific questions. As a researcher, you need to know the answer to the following questions concerning the research topic you are *considering* for investigation:

- Has the work already been done?
- Who are the experts in the field?
- What are the main theoretical perspectives?
- What are the common research methods in the topic?
- What are the main problems in researching the topic?
- Are there any major controversies in this topic area?
- Is the topic open to hypothesis testing?
- Is the topic a trivial one?

The only way you can satisfactorily answer these questions is by reading as much as you can on research which is directly related to your research topic, research which is indirectly related to it and research which *may* be related to it. In the case of the last one, you can only know this by reading material you think might be relevant. The easiest way to identify such material is through the title and the abstract. Let us go through each of the questions posed above.

Has the Work Already Been Done?

Obviously, if this is the case then you need to consider changing your topic or changing its focus. In most business-related research, it is very rare that a specific set of research questions or a specific hypothesis has already been addressed in your specific topic area. This is more common in science research. Nevertheless, it is still important to check that a piece of research already published is not so close to yours that undertaking your research would not be worthwhile. Clearly, this means that the literature review and dealing with the first question above are factors that need to start immediately after you have identified a possible research topic.

Who Are the Experts in the Field?

There are many areas of business research where it is very difficult to identify any experts, but there are areas where it is not. For example, in the field of macroeconomics it would be very difficult to avoid the work of J.M. Keynes because this author has published widely on this topic and proposed an important theoretical model of how an economy works. In the field of 'Bureaucracy in Organisations', it is crucial that the work of Max Weber is consulted as well as critiques of his work.

If your topic was strongly linked to business forecasting, you would need to consult the publications of Makradakis and/or published material which reports on his work. It should be clear to you now that in many areas of business research there are indeed experts on specific topics—so how do you identify them? In fact, this is relatively easy—you could consult the Social Science or Humanities Citation Index by topic area (on the web) and this will show up the names of authors in that area. If a name appears several times, then you can be reasonably certain that the author is very active in that research topic.

Alternatively, you could find a published article related to your topic in an academic journal and consult the reference list at the end of it—again it is often the case that the same name is referenced more than once. This is also a good way of establishing a 'road map' for your literature review—by following up references used in books, journals, academic working papers or even in electronic websites. This is because most of the references will be strongly related to the topic of the article.

What Are the Main Theoretical Perspectives?

Whatever the research topic you have identified, you need to construct a conceptual framework within which you will study the topic. This is critical to the successful implementation of the Research Cycle discussed in Chapter 3. Without theory, it is almost impossible to interpret data. For example, suppose I collected data on the number of washing machines purchased in a single city in a single year—this would be relatively easy to do, but once I had the data what would I do with it? I could graph it, I could apply all types of statistical analyses to it and I could write several paragraphs describing it. However, I could not even begin to try and EXPLAIN it! This is because I have *no theory* of the consumer decision to buy a washing machine. In fact, there is no point in graphing this data, in analysing it or in describing it unless I am a pure inductivist (see Chapter 2) and have no idea whatsoever of the reasons behind washing machine purchase. Even worse, why would I collect such data in the first place if I have no understanding of why I am collecting it, if I have no *a priori* reasons to collect it and if I have no hypothesis I wish to test?

The key point here is that *I need* a theory in order to inform me what kind of data I need in order to answer the research questions I have already set myself—in other words following the modern deductivist methodology of research. Where do I find this theory? The answer to this question is easy—you will find it in the literature related to your research topic. You will also find critiques of theory there and alternative theories of the same social behaviour. It is critical that you are aware of all theories pertaining to your research topic, their strengths and their weaknesses. In fact, it is often the case that the theoretical knowledge of a particular topic reveals gaps in our understanding of that topic—this often enables you to identify more important research questions.

What Are the Common Research Methods in the Topic?

In reading the published academic literature on or closely related to your research topic, it is very important to try to identify the common characteristics of HOW the research was carried out. There are a number of aspects of this that need to be clearly understood:

1. Is the published research in this topic mainly of a qualitative or a quantitative nature?
2. If qualitative, what is its most common basis?
3. If quantitative, what is its most common basis?
4. Is the published research usually a mix of qualitative and quantitative methods?

It is relatively easy to answer the first question. An initial review of a handful of published papers on the topic will quickly reveal the nature of the research approach taken to the topic—these will either include a lot of mathematical analysis or statistical data or be dominated by textual analysis. The latter usually indicates a qualitative method being used. If this is the case, then we can move to the next question: are the qualitative methods mainly of the case study type, ethnographic, content analysis, grounded theory or some other generally recognised qualitative research method? It is your judgement as to what appears to be the most common basis for research in this topic. A detailed discussion of both quantitative and qualitative methods is provided in later chapters.

If you find that the handful of papers are dominated by either mathematical or statistical analysis, then you need to identify the nature of these. In business research, quantitative methods tend to be dominated by statistical analysis. You need to be clear on the most common approach taken here—is it simple descriptive statistics, non-parametric methods, parametric methods, multivariate methods or another generally accepted statistical approach? The answers to questions 1 and 2 above will inform you as to *how* you should be formulating your own research questions as per the Research Cycle we considered in Chapter 3. If, as is often the case, the published research is a mix of qualitative and quantitative methods, then you need to decide where you will put the emphasis—and this will very much depend upon how you have formulated your specific research questions or how you hope to formulate them.

You also need to consider if the published research derives its data from secondary sources mainly or primary sources mainly—this will also inform you as to how and where you should be seeking your own data. Also consider what type of data is most commonly used in the topic: cross-sectional, time-series or categorical, for example.

In addition, what is the typical (if any) source of this data—focus group, case study, survey or published sources base? It is also useful at this early stage of the literature review to determine if the data is being used or is being tested against a theory or a theoretical model or if it is very much exploratory, that is, being applied within a predominantly inductivist framework.

What Are the Main Problems in Researching the Topic?

Without a review of the literature, you cannot possibly know this. There may be many problems and several common problems in researching your research topic. One of the most common problems is data—for example, does it exist? If it does, can you get access to it? Is it in a format which is easily manipulated? Is it trustworthy? How old is it? Have the variables in the data been measured correctly and consistently? All of these questions are important. In addition, is the topic one for which primary data is essential? If so, can it be collected in a reasonable time and at low cost? Can it be collected at all if the topic is personally, socially or politically sensitive? Another problem which can arise is the absence of a clear theoretical framework in the published research. This may be a topic which has little or very weak theoretical underpinning and understanding—if so, then how can you test a hypothesis or attempt to answer your research questions? This needs to be considered.

Finally, the topic may be one in which there is serious disagreement over how to research it at all—in other words, there is a real concern surrounding the value and appropriateness of any of the research methods used to investigate the topic. If this is the case, you should seriously consider if this is a topic suitable for a dissertation.

Are There Major Controversies in This Topic Area?

The controversies may be methodological, theoretical or empirical. It is one of the skills of literature reviewing to determine what the controversies actually are. This is also a very fruitful source for deriving your own research questions since the gaps in our understanding of the topic will be fairly clear from the debates and controversies you find in the literature.

Is the Topic Open to Hypothesis Testing?

Not all research questions can be put in the format of a hypothesis. A hypothesis needs to be clear, unambiguous, focused and testable using an accepted statistical method. If the research topic is mainly investigated using qualitative methods, then it may be very difficult to generate testable hypotheses. However, a good piece of research does not need to contain testable hypotheses—it could contain a set of very focused research questions (but not too many) or a set of clear propositions (derived from theory). In either case, you can still 'test' these using a logical and discursive analysis and/or statistical methods appropriate to the type of data being used.

Is the Topic a Trivial One?

If any one person or any one organisation has a 'problem', then it will never be trivial to them. However, research should be concerning itself with problems which are more significant and with the potential for the results to be generalisable to at least a group, an area, a sector of industry or any other aggregation. A good indication of a research topic which is considered trivial is when you can find little or no trace of it in the literature. Of course, this could be because no one has ever thought about it but it is usually because they have and immediately dismissed it as of no interest to anyone else, of no value to society and has no potential to be developed further. It is very much a matter of judgement.

4.3 WHAT SHOULD THE LITERATURE REVIEW DO?

Apart from providing the answers to the above questions, the review should also achieve the following:

- It should enable you to sharpen and focus your initial research questions or even suggest new research questions.
- It should provide you with a wide and in-depth knowledge of the theoretical, empirical and methodological issues within your chosen research topic.

- It should provide a 'bridge' between your research questions and your research findings.
- It should enable you to speak with authority on your research topic and the wider subject area.
- It should enable you to compare your research methods, theoretical framework and findings with work already done.
- It should enable you to set the scope and range of your research topic.

Quite clearly, the literature review is the pivotal element of a research project. It connects your planned work to previous work, it connects your specific topic to the wider subject area and it connects your specific findings to the findings of others. It is very much an inclusive activity in the sense that, if undertaken properly, you become a part of the academic community who can speak and write with confidence and authority on a specific research problem.

Linking to Your Research Questions

Most research projects begin with no more than initial ideas—very often unfocused and based mainly on personal interest. This is normal; however, the literature review will enable you to do a number of things in relation to any initial research questions you may have:

- It will allow you to discard ideas considered trivial in the literature.
- It will enable you to discard any questions the literature shows are extremely difficult to deal with.
- It will allow you to frame your research questions in the context of the main theories present in the literature.
- It will enable you to identify research questions which potentially can fill a gap in knowledge identified in the literature.

The literature review thus helps you narrow down the focus of your research and to be much more precise in framing the research questions or hypotheses which interest you.

Linking to Your Research Methods

As discussed earlier, it is very important to be aware of the main research methods employed in your chosen research topic. The literature review will help you identify *how* you should be designing your research project in order to answer the research questions you have posed. An understanding of what works well and what does not work well in terms of method is crucial to identifying an efficient and effective research method of your own. It may not seem like it but the literature review will actually save your time—because it enables you to avoid mistakes and to avoid reinventing the wheel! An understanding of the range of research methods employed in a particular research topic also enables you to identify the Limits of Validity of any findings because you will have understood what a particular method can achieve and what it cannot achieve compared with alternative methods. This is very important because, as you know from earlier chapters, any research finding is itself dependent on how the research was carried out and can never be assumed to be the 'last word' on the subject. Thus, a good understanding of methods in a particular research area will enable you to identify areas for future research and to be realistically modest in your assessment of your own findings.

Linking to Your Theoretical Framework

This is arguably the most important part of any research project and the most important function of the literature review. As you know from earlier chapters, data comes and goes, the relationship between variables changes, society changes and therefore the primary function of academic research is to continually move theory forward—because in the end, it is all that we really have in terms of our 'knowledge' of the world. A research project that heavily relies on the description of data trends, the description of constructs and the description of behaviour is extremely limited in what it can contribute to knowledge. As you have already seen in Chapter 2, the nature and purpose of the research project will largely determine how the work is undertaken, what research questions are feasible and which, if any, theoretical framework can be used.

Unless the data, the constructs and the 'behaviour' are interpreted in terms of theory, the research work itself will be no more than a commentary on a specific topic. For example, most business students have encountered the concept of the 'demand curve' from economics and know that as prices fall, demand rises, and we can trace this by moving the price line along the demand curve. However, this is merely a description of what happens when prices fall—it is not the explanation. For the explanation, we need to consult the utility theory. There is a clear difference between description and explanation and it is the latter that can only be provided by theory.

The literature review will also allow you to understand competing theories of social/business behaviour and to be aware of the weaknesses of these theories. It will also enable you to attempt to integrate different ideas from different theories in order to construct your own conceptual framework and to link this to the research questions you are asking as shown in Figure 4.1.

To the above process, you can add different datasets, related issues and sub-themes of the research topic. The key element here is that there must be a theoretical framework

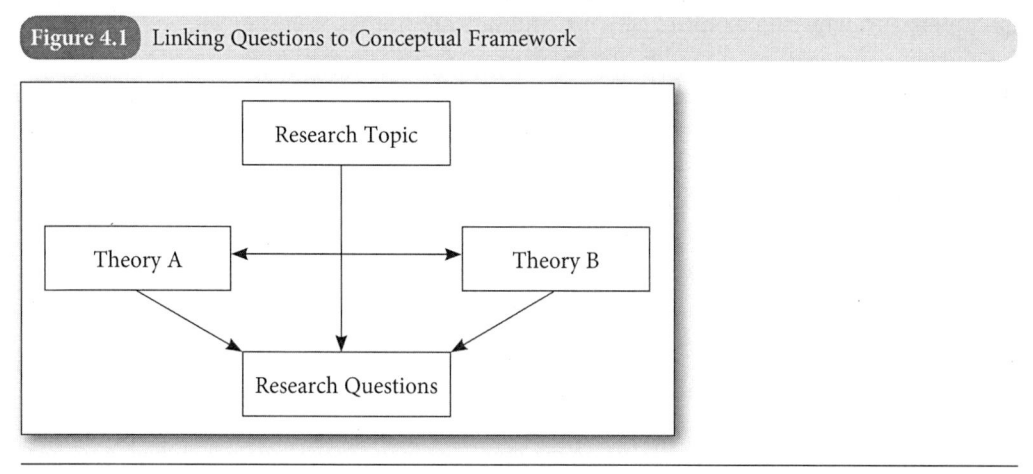

Figure 4.1 Linking Questions to Conceptual Framework

```
            Research Topic
                 │
    ┌────────────┼────────────┐
    ▼            ▼            ▼
 Theory A ◄──────────────► Theory B
    │            │            │
    └────────────┼────────────┘
                 ▼
          Research Questions
```

SOURCE: Authors' own.

in order to allow you to interpret your results and to identify what has been achieved and what has not been achieved. This will also enable you to question whether the weaknesses in your own research are theoretical, methodological or related to the data you have used. Finally, you need to interpret the results of your research and the literature review; the theory will help you to do this.

Linking to Your Research Findings

You need to be in a position to be able to compare your findings with the findings in previous research. Obviously, you can only do this if you are aware of the findings in the literature. The theoretical framework allows you to interpret the findings while previous research allows you to compare these with the work of others. In addition, you should be evaluating your findings in the context of the research questions you have identified and subsequently sharpened in the light of your understanding of the literature. In order to link your findings to your literature review, you need to consider a set of questions, some of which you may be able to answer and some of which it may not be possible to answer; nevertheless, they must still be explicitly considered and a discussion of them reported in your work. Specifically you need to ask the following questions:

- Which research questions have been satisfactorily dealt with?
- Which have not been satisfactorily dealt with, and why?
- Which have not been answerable at all, and why?

You should also be identifying findings you know to be consistent with the literature and those you know are not consistent with the literature. In addition, you need to be able to explain why one or more of your findings are not what you expected (given you are *a priori* reasoning) and to identify the source of this—it may be data weakness, sampling problems, theoretical weakness or it could well be that one of your specific research questions was not framed correctly or was in fact irrelevant. It is part of the research and evaluative process to work this out and properly report it. Finally, you need to consider the Limits of Validity of your findings, the extent to which they can be generalised and the extent to which you can claim a degree of reliability of the findings. Again, all of this can be done with reference to your literature review.

4.4 TYPES OF LITERATURE REVIEW

There are different types of literature review that can be undertaken, depending on the purpose of the research. The main types of literature review are as follows:

- An evaluative review
- An exploratory review
- An instrumental review

These are not mutually exclusive and will often be mixed together; however, in the case of academic research it is the second one that is the most common.

Evaluative Review

This focuses on providing a discussion of the literature in terms of its coverage and contribution to knowledge in a particular area. An extreme example of this type of review is meta-analysis, which provides a comprehensive commentary on a very large number of research projects focused on a specific topic. It is often used to directly compare research findings from these projects when the findings are directly comparable—for example, in measuring reliability coefficients, regression coefficients and artificial constructs are defined in the same way but applied in different projects. The field of econometrics is often typified by literature reviews of this type.

Exploratory Review

This literature review seeks to find out what actually exists in the academic literature in terms of theory, empirical evidence and research methods as they pertain to a specific research topic and its related wider subject area. It is also used to sharpen, focus and identify research questions that remain unanswered in the specific topic. The key to conducting a review of this type is to remain focused on the field of study and not allow yourself to be taken into other directions just because they may be interesting. It is not as important here to provide a comprehensive review of the literature (as in meta-analysis), but it is much more important to focus on the specific area of the research topic.

The review should be seen as being informative to the researcher and providing the researcher with clearer ideas on the common theories, methods and types of data analysis conducted in this topic. It is also critical in this type of review that the literature is properly cited and a proper bibliography is presented. This is to enable other researchers and readers to follow up aspects of the work they find especially interesting. In most academic works, the most common referencing system is the Harvard system. This is very much the basis of an academic literature review designed to inform and create a 'path' between previous research and current research.

Instrumental Review

This is where the literature is used exclusively as a source of information on *how* to conduct a piece of research on a highly specific research problem. It is not designed to identify the state of current knowledge in an area but to identify the best way to carry out a research project without reinventing the wheel and without incurring unnecessary and avoidable costs. This is the type of review which would be typically carried out in-house by company employees who are tasked to 'solve' an urgent or expected business problem. This type of review will also be less concerned with properly citing the literature unlike the two above.

4.5 SOME GENERAL POINTS IN LITERATURE REVIEWING

Reviewing academic literature is not the same as just reading it! You need to think about the ideas, the research methods, how data was collected and how the findings have been interpreted. This is what we mean by Critical Reading and you will study this in some detail in a later chapter. In the meantime, here are some questions you should keep in mind when studying (not just reading) academic literature.

For any given piece of work:

- Is there a theoretical framework?
- If so, what is it and how does it fit into this topic?
- Does the work provide links to other work in the topic?

- Is there an empirical aspect to the work?
- If so, what is its basis?
- Does the work relate to a specific social group?
- Does it relate to a particular place?
- How applicable might it be outside the latter two?
- How old is the work?
- Is it still valid?

In addition to asking these questions, you also need to summarise the literature review. This should be done in the following terms:

- What does the previous research tell us about this topic?
- What does it not tell us?
- What are the key weaknesses in terms of theory, methods and data?

This is the 'end' of your literature review and you now have a platform from which to launch your own research, interpret the findings and evaluate what you have achieved in comparison with the literature. Of course, as explained at the beginning of this chapter, the literature review is never really complete simply because there is always work on-going, which you will be unaware of, there is often too much literature to be covered in a specified period of time and there will be literature on your topic which is already in the process of being published but is not yet published. This is entirely normal and nothing to be worried about.

A good review should demonstrate familiarity with the topic, show the path of prior research and how it is linked to the current project. To do this effectively, the review should be written in a critical and reflective style. One should not simply accept something because it is written; we should use our judgement to decide where it is good or where it is poor. Being critical does not mean simply picking holes in an argument; good ideas and well-developed arguments should be praised too.

4.6 OBTAINING LITERATURE SOURCES

There are many literature sources including journals, books, reports, abstracts and electronic websites. Searching for the appropriate literature can be very time consuming and you need to be very specific when using library search engines, Internet search engines and other databases.

Once an appropriate article is obtained, there are various ways of reading for research purposes. In doing the reading for the review, one must consider the credibility of the article. Articles published in refereed journals tend to be the most reliable. Articles obtained from the World Wide Web (www) have to be treated with a great deal of caution, although there is good material on the web. If you plan to use the Internet as a major source of research material, the most reliable sites are those of academic departments in universities. For example, if you are researching a topic in the area of financial innovation or marketing communications, a good start point would be a Department of Finance in the first case and a Department of Marketing in the second case. Most universities have academic departments where the departmental website will contain staff research papers, staff working papers and sometimes staff publications—in most cases, these can be downloaded in full and at no charge.

Another source of reliable material is government departments—they often put full reports and analyses of specific topics on their websites and again often at no charge. It is not a good idea just to enter the name of a topic into an Internet search engine because it will find material you cannot be certain is free from bias or has been through a refereeing process. Material supplied by academic departments and government departments will have been refereeing one to another and editing. These are far more reliable.

However, there is no real substitute for spending a considerable amount of time in libraries, playing detective and tracing articles cited as references to articles you have read. You will not be able to read everything, so be selective. Reading abstracts helps in this. Remember that photocopying articles, although reassuring, is not a substitute for reading them. When you make photocopies, always ensure that you copy the references at the end of the chapter.

On reading an article, it is useful to make notes and record accessing details on a card or computer file. Do this recording at the time, as failure to do so can add a considerable amount of time to your writing of the literature review and to your construction of the bibliography.

4.7 SEARCHING THE LITERATURE

Literature search involves a systematic and methodical search of published sources of information to identify items relevant to a particular need. The 'literature' involved may be in the form of books, journal articles, videos, cassette tapes, conference papers, reports, theses, patents, standards or other types of information source.

Why Do a Search?

- To help in topic selection—to assess a topic's novelty, originality and feasibility
- To discover information that you can use in your actual project—it is vital to show that you have located, understood and assimilated previous work in the field

- To provide knowledge of the subject area in general, for background and contextual information—a search will make you aware of the structure of information in a particular subject area

Planning the Search

1. Plan your information search. Spend some time thinking about what you really want to find information on. This may involve breaking your topic down into several separate information searches.
2. Consider the following aspects of your search:

   - *Scope*—is the information you require—a core element of your topic, is it background (a paragraph in your introduction) or tangential to your central topic?
   - *Timescale*—how far back is the information of relevance? Must it be recent? Published in the last five years? Or historical? If looking to forecast, you may need to look back over a long period to assess statistical trends for example.
   - *Range*—do you need local information only? Regional? National or international? If a variety, assess the part each type will play in your project.
   - Set yourself parameters or limits to the search if you can—but keep it flexible, as your strategy may change as the search develops.

3. Most importantly, select the *key concepts* your search will involve. From these concepts, generate a number of subjects or *keywords*. These are the words you will actually look up in various information tools.

   To help you generate keywords, look through textbooks, encyclopaedias, handbooks, etc., or scan some current issues of journals in the appropriate field. Brainstorming is an excellent way to identify keywords, especially in groups. Integrate new keywords discovered as your search progresses. When keywords are handy and appropriate, they will give your search some element of strategy. Think in *broad* terms that encompass your topic, *narrower* terms that are more specific, *related* terms that will enable you to enrich your search and *synonyms* or *alternative* terms to make your search comprehensive. When thinking of synonyms, be aware of global terms that may be used, and check out each in the indexes you select.

   Think about the *type* of information you are seeking. This may help you identify the most likely tools to use. Is it general theory (textbooks), current analysis and comment (newspapers and journals), names of contacts (directories), statistical, governmental, legal, technical or bibliographical?

Doing the Search

1. Consult library catalogues, general subject guides, bibliographies, etc., to establish which indexing tools will be the most useful for your topic. Ask library staff for assistance in selecting the most appropriate tools.
2. Using the selected sources' search through the subject indexes to trace relevant articles. If you find new or relevant keywords, add them to your search. Locate as many references as possible on your topic. If you cannot find many, it may be because the keyword used is too narrow—broaden your search using more general terms. If you find too many articles, try more specific keywords. Use your synonyms and amend your search in the light of experience. If you cannot find much at all, it may be that you are using an inappropriate index or, as explained above, it may be because the topic itself is not considered important and therefore there is very little work on it. Another explanation could be that the topic you have chosen is so new that there has been no time for published research to appear yet. An example of this is the area of the 3G mobile-phone service—it has been launched in very few places only recently and economic, marketing and social research into this topic has yet to appear in significant numbers.
3. Be methodical and patient. Information search can be frustrating, and there are times when you will find either too much or too little.
4. When you do find relevant articles or books, *note down the full bibliographic reference*. This will save you a great deal of frustration later on when you come to compile your bibliography. If you are doing an in-depth project or dissertation, you may wish to compile a database of your references with a list of keywords describing the contents of the work. This can be compiled manually or on computer. You may also wish to add your own notes, detailing how useful the article was, where you obtained it and what it has given to your literature review and to your research project in general.
5. Depending on the results, you may wish to extend your search by accessing information sources outside the library. This may take the form of an online search accessing remote, computerised stores of information. You may also wish to use other information sources external to the university such as specialist libraries or information chapters, personal contacts, etc. Before you do this, it is essential that you have a clear idea of what you are looking for, and that your planning has been as thorough as possible; otherwise, you will waste much time and effort and results will be variable.

4.8 ASSESSING THE QUALITY OF LITERATURE

It is not easy to assess if a piece of published work is of high, medium or low quality until you actually read it and are able to compare it with other work you have read. However, there are a number of 'tests' you can use to give yourself a much clearer idea of what represents good and not-so-good research work. When reading a piece of published work from an academic journal, a newspaper, a textbook, a government report, a company report, a popular magazine or from the Internet, you can ask yourself the following questions:

- Is it clear what organisation is responsible for the contents of the work?
- Is there a way of verifying the legitimacy of this organisation? That is, is there a phone number or postal address to contact for more information?
- Is there a statement that the content has official approval of the organisation?
- Is there a statement giving the organisation's name as copyright holder?
- Do you know who wrote the article and his or her qualifications for writing on this topic?
- Is it clear who is ultimately responsible for the content of the material?
- Are the sources for any factual information clearly listed so that they can be verified in another source?
- Is the information free of grammatical, spelling and other typographical errors? (These kinds of errors not only indicate a lack of quality control, but can actually produce inaccuracies in information.)
- Are there editors monitoring the accuracy of the information being published?
- If there are charts and/or graphs containing statistical data, are the charts and/or graphs clearly labelled and easy to read?
- If material is presented in graphs and/or charts, is it clearly stated when the data was gathered?
- Is it clear when the work was published?
- If an academic journal article, when was it submitted to the journal?
- When was it accepted in the revised form?

It should be clear to you by now that the safest source of material for research is academic journals and the websites of academic departments. The first is where a full and proper academic refereeing and editing process is guaranteed and the second is where a similar process is very likely to have occurred, and if not, the authors' own academic credibility is usually sufficient to ensure that this material is of a very high standard.

4.9 AN EXAMPLE OF A LITERATURE REVIEW

Below is an extract from an article published in an academic journal in 2002. The topic of the article is Financial Globalisation and only its literature review section is reproduced here. You should read this, followed by a different version of the same literature review. In reading the different (second) version, you should consider its difference from the first version. We provide a commentary on both, which hopefully you will be able to compare with your own assessment of the two versions and the differences between them.

Of course, not all the aspects of a literature review (discussed above) can be covered in a single example but the key elements of what needs to be present are contained in this example.

We also provide a second example of a literature review, again from a published paper and again in two versions. However, in this case we provide no commentary on the review and leave it to you to assess what the key differences are—this may also be used as an exercise at your tutorial or workshop or in discussion with your fellow students.

Example 1—Extract from:

Financial Globalisation: Is It a Threat to Key Central Bank Functions? The Case of Mauritius

John Adams*

ABSTRACT

There has been much discussion over recent years of the likely impact of 'financial globalisation' on the financial services sector specifically and on the stability of national economies. This Paper examines how and to what extent the 'threats' from financial globalisation manifest themselves in relation to the functions of regulation and supervision carried out by central banks. A theoretical perspective on these issues is put forward followed by an analysis of the specific case of a small island economy which is embracing financial liberalisation and competition, Mauritius.

Published in *The International Journal of Financial Regulation and Compliance,* Vol. 10, No. 1, February 2002.

Received March 2001

In revised form, September 2001

*School of Accounting and Economics, Napier University, Edinburgh, EH14 1DJ
TEL: 0131-455-4308 Email for correspondence: j.adams@napier.ac.uk

(Version 1)
I. Introduction

The development of global financial transactions has been a key feature of international economic activity since WWII and particularly since the 1970s. In more recent years total international bank lending has been expanding by approximately eight per cent per annum (BIS Annual Reports 1997(b) and 1999). However, the expansion in net international capital flows has not been without interruption. It has been clearly demonstrated that this process has followed a pronounced cyclical pattern. A rapid expansion in the 1880s was followed by rapid decline (1890s), expansion (1910s), slow decline to the 1930s, expansion (1940s) and slow decline to the early 1960s. Since then international capital flows have been subject to shorter cyclical fluctuations but the long-term trend has been one of steady expansion between 1970 and 1980, followed by rapid expansion since the early 1980s. A similar pattern can be observed in relation to total foreign exchange market turnover, as would be expected.

Allied with these developments has been what can fairly be described as an explosion in financial innovation and the advent of a previously unheard of level of liberalisation in both domestic and international financial markets and services. The key features of these developments have been an extraordinary growth in the 'off-balance sheet' activities of financial institutions and in 'over-the-counter' (OTC) transactions in options, swaps, and other financial derivatives. These are very important developments for the management of national economies, particularly in the area of monetary policy. This is because such innovations have effectively surpassed existing national and international regulatory and supervisory frameworks in many cases. It is therefore becoming increasingly difficult for central banks and other monetary authorities to undertake the primary functions of financial market stabilisation and price stability within their domestic economies to pursue past policy regimes typified by the financial repression paradigm identified by McKinnon.

It has recently been argued that these developments have in fact made the traditional regulatory and supervisory functions of central banks almost impossible. In other words, the risk of a financial 'crisis' both domestically and internationally has increased and the speed of transmission (contagion) of such crises has also increased. The challenge facing central banks and the international financial market system is how to deal with the increased risks which are now embedded within an increasingly interdependent but globalising financial market system.

This is clearly an issue for all central banks. However, this paper is confined to the particular case of Mauritius and how the central bank might respond to such a challenge.

The paper is structured as follows: in Section II, the concept of financial globalisation is discussed. This provides a useful theoretical framework within which the case of Mauritius can be analysed (Section III) and in Section IV of the paper, a number of key issues are identified with respect to 'what can be done'. Section V concludes the paper.

II. Financial Market Globalisation—What Is It?

Perhaps the clearest answer to this question is first to describe what financial globalisation is definitely not. It is not simply the expansion of net or gross international capital flows; it is not the expansion of individual economies' external financial transactions per se and nor is it the increasing entry of large institutions into non-home-based markets. These are all aspects of the *internationalisation* of capital and, as we already know, are subject to significant cyclical fluctuations. Were these the key elements of financial globalisation it would be irrefutably logical to talk of periods of 'de-globalisation' such as in the 1930s and 1950s. In addition, there remains a strong tendency for investors and their portfolios to retain a very strong 'home-bias' and for wealth in all its forms to be predominantly held within investors' home countries. This home or domestic bias is likely to decline with time, but it is an indication that the completely 'globalised' financial world is still some way off.

It seems clear, therefore, that cyclical fluctuations in cross-border capital transactions can neither conceptually or empirically form the basis of what is effectively a relatively new and even 'populist' concept. Instead, we need a clear and unambiguous theoretical rationale, which is capable of providing deeper insights into a process which we *believe* is under way but as yet do not fully understand. This will enable further analysis of the phenomenon and provide a better understanding of its implications for national economies and for the role of central banks in the future. As Shirikawa et al argue:

> "… globalisation refers to [a situation] where each country's economy, including its financial markets, becomes increasingly integrated resulting in development towards *a single world market*."

In other words, financial market globalisation will not and cannot proceed in the absence of the globalisation of all production relations, including labour itself. Using the definition given above, it is possible to delineate stages of the globalisation 'process' and to categorise what these stages mean in terms of both the 'real' economy and the financial markets. Consider Figure 1.

Commentary

The first thing worth pointing out here is that the literature review above contains 845 words—this is for an article which is 4,200 words long. In other words, the literature review is a significant part of the whole article, nearly 20 per cent. This is quite normal in research work and, as a general rule, you will find that for a Masters Dissertation of between 12 and 15 thousand words the literature review will typically be between 3 and 4 thousand words. Let us consider the short literature review above. The first thing to note is that where a fact or a theory is being referred to, it has been referenced. Note too that the extract contains a quotation—not only is the quotation referenced but the exact page number of where the quotation is located is also provided. As explained above, it is important that this is done, hence the importance of always recording key elements of what you read as you read.

Second, the identity and affiliation of the author is clearly shown. Third, the date of publication is clearly shown and fourth, the date of the manuscript (first received by the Journal and subsequently received in revised form) is also clearly shown. Some academic journals do not always give this information. However, where it is given, you should consider the gap between submission date and the final publication date.

The next thing to notice is that this is an example of an exploratory literature review—it attempts to assess what we already know about financial globalisation from the literature and then goes on to set out how this knowledge is going to be applied to the topic of interest—the case of the Mauritian central bank. The review also attempts to give the reader a clear explanation of what financial globalisation is not and explains this in terms of a very brief analysis of historical trends. This assessment of directly relevant literature is then used to spell out the basis of the Paper and what the rest of the Paper will do. In other words, a clear path is constructed, connecting the literature to the present study.

Notice too that the literature review is by no means comprehensive—it is highly selective and focused on centrally relevant material. A much more comprehensive review of the work in this topic could have been done but that would then make the purpose of the research and of the Paper quite different. Hence the importance of clearly setting out your research questions and of carefully constructing your research design.

Now let us consider a different version of the same extract and try to identify what is different and, more importantly, what might be wrong, if anything, with this different version.

(Version 2)

I. Introduction

The development of global financial transactions has been a key feature of international economic activity since WWII and particularly since the 1970s. In more recent years, total international bank lending has been expanding by approximately eight per cent per annum (BIS Annual Reports). However, the expansion in net international capital flows has not been without interruption. It has been clearly demonstrated that this process has followed a pronounced cyclical pattern. A rapid expansion in the 1880s was followed by rapid decline (1890s), expansion (1910s), slow decline to the 1930s, expansion (1940s) and slow decline to the early 1960s. Since then international capital flows have been subject to shorter cyclical fluctuations but the long-term trend has been one of steady expansion between 1970 and 1980.

Allied with these developments has been what can fairly be described as an explosion in financial innovation and the advent of a previously unheard of level of liberalisation in both domestic and international financial markets and services. The key features of these developments have been an extraordinary growth in the 'off-balance sheet' activities of financial institutions and in 'over-the-counter' (OTC) transactions in options, swaps, and other financial derivatives. These are very important developments for the management of national economies, particularly in the area of monetary policy. This is because such innovations have effectively surpassed existing national and international regulatory and supervisory frameworks in many cases. It is, therefore, becoming increasingly difficult for central banks and other monetary authorities to undertake the primary functions of financial market stabilisation and price stability within their domestic economies to pursue past policy regimes typified by the financial repression paradigm.

It has recently been argued that these developments have in fact made the traditional regulatory and supervisory functions of central banks almost impossible. In other words, the risk of a financial 'crisis' both domestically and internationally has increased and the speed of transmission (contagion) of such crises has also increased. The challenge facing central banks and the international financial market system is how to deal with the increased risks which that are now embedded within an increasingly interdependent but globalising financial market system. This is clearly an issue for all central banks

II. Financial Market Globalisation—What Is It?

Perhaps the clearest answer to this question is first to describe what financial globalisation is definitely not. It is not simply the expansion of net or gross international capital flows; it is not the expansion of individual economies' external financial transactions per se and nor is it the increasing entry of large institutions into non-home-based markets. These are all aspects of the *internationalisation* of capital and, as we already know, are subject to significant cyclical fluctuations. Were these the key elements of financial globalisation, it would be irrefutably logical to talk of periods of 'de-globalisation' such as in the 1930s and 1950s. In addition, there remains a strong tendency for investors and their portfolios to retain a very strong 'home-bias'. This home or domestic bias is likely to decline with time, but it is an indication that the completely 'globalised' financial world is still some way off.

It seems clear, therefore, that cyclical fluctuations in cross-border capital transactions can neither conceptually nor empirically form the basis of what is effectively a relatively new and even 'populist' concept. Instead, we need a clear and unambiguous theoretical rationale, which is capable of providing deeper insights into a process which we *believe* is under way but as yet do not fully understand. This will enable further analysis of the phenomenon and provide a better understanding of its implications for national economies and for the role of central banks in the future. As Shirakawa et al argue:

> " ... globalisation refers to [a situation] where each country's economy, including its financial markets, becomes increasingly integrated resulting in development towards *a single world market*."

In other words, financial market globalisation will not and cannot proceed in the absence of the globalisation of all production relations, including labour itself. Using the definition given above, it is possible to delineate stages of the globalisation 'process' and to categorise what these stages mean in terms of both the 'real' economy and the financial markets. Consider Figure 1.

Commentary

So, what can we say about version 2? Hopefully, you will have identified a number of discrepancies and errors and omissions of explanation. First, the referencing: in the first reference all that is supplied is 'BIS Annual Reports'—no information on which annual reports the discussion refers to is given. This, therefore, makes it very difficult for the reader or other researchers to check the facts for themselves. In the second reference the initial of the first author is given but that of the second is not—this is inconsistent and should be avoided.

The end of the first paragraph stops the discussion of 'trends' at 1980—yet the Paper is published in 2002; hence, the discussion of the 'trends' is clearly incomplete.

At the end of paragraph 2 the term 'financial repression paradigm' is introduced. This is a body of theory and yet there is no reference to indicate the source of the theory to allow a reader to follow it up and learn more about 'financial repression'. It is extremely important that when using what is clearly a theoretical concept it should either be explained or be given a clear reference to which the reader can be directed. In paragraph 3 something is identified as 'clearly an issue for all central banks'—but this Paper is not about all central banks, it is focused on a particular country's central bank and this is not made clear. Thus, an opportunity to guide the reader towards the real focus of the research and the Paper has been lost.

In paragraph 4 the term 'home-bias' is introduced but there is no attempt to provide an explanation of what this is. It is left to the reader to 'guess' what it might be. This is very poor practice in literature review and must be avoided—just because you know what something means, do not assume that the readers will. In short, the introduction of concepts, 'jargon' or any other category of information should always be explained, ideally as briefly as possible. With a Masters Dissertation (or a journal article) you are not writing a textbook; therefore, you need to be precise, to the point and economical with language. The quotation given in the review is referenced but the page number of its source is not given. This is also a poor practice and should be avoided.

Finally, the review gives no indication of what is still to come in the rest of the Paper. There is no identification of what the reader can expect because there is no structure given, linking the review to the rest of the Paper. It is essential that you provide forward linkages to what is still to come and, later in the Paper or Dissertation, backward linkages to what you have already discussed.

We hope that you were able to identify at least some of the differences between Versions 1 and 2 and able to see what is wrong with Version 2. It is also useful at this point to note that the commentary for Version 2 is also an example of Critical Reading. In order to write this commentary, simply reading Version 2 was not enough—it had to be studied carefully in order that the 'critique' of it is fair, accurate and sensible.

Example 2—Extract from:

Air Passenger Growth Forecasts for the United Kingdom: The Potential Threat of the Policy Alternatives to Scottish Air Travel

John Adams* and Robert Raeside

Published in *The Fraser of Allander Quarterly Economic Commentary, December 2001*.
Received August 2001

In revised form, October 2001

(Version 1)

1. Introduction

The DETR has requested responses to its consultation document on the future of air transport services in the United Kingdom. A number of alternative policy responses are set out, which are rooted in the DETR's most recent forecasts for passenger growth. The forecasts of demand for air travel are and will continue to be at the core of these policies. However, it is argued in this paper that potential policy responses to the forecasts run the distinct risk of being contradictory, socially divisive and a threat to the continued expansion of air travel services in Scotland and also in all regions outwith London.

In the case of Scotland, the 'demand constraint' option runs counter to the recent argument from the Secretary of State for Scotland that the country is still poorly served in terms of air transport. This article is concerned with setting out the context of the issues, the efficacy of the forecasts upon which the 'policy options' are being mooted and the implications for Scottish air travel if several of these options are taken up. First, it is useful to present some background information on UK air passenger growth.

Between 1974 and 1999 the number of air passengers travelling into and out of the United Kingdom (UK) has increased from 49 million to 171 million, an increase of almost 250 per cent. The average annual increase in passenger numbers has been slowing down since the 1960s (Table 1):

*Corresponding author
School of Accounting and Economics
Napier University Business School
Craiglockhart, Edinburgh EH14 1DJ
Tel: 0131-455-4308
Email:j.adams@napier.ac.uk

Table 1	Annual Average Growth Rate in Passenger Numbers				
Decade	1960–69	1970–79	1980–89	1990–99	2000–2010
Growth Rate %	14	7.3	5.7	5.1	4.3

SOURCE: DETR.

The declining rate of growth reflects a consistent movement towards market maturity in passenger air travel for the UK. The decline is expected to continue in the present decade towards an annual average growth rate of 4.3 per cent and is expected to stabilise around this rate for the foreseeable future. This is consistent with forecasts suggesting a doubling in passenger air traffic over the next ten years and a near doubling in the last ten years on a *global* basis. Although the expected lower rate of growth for the UK in the next ten years is considerably less than that recorded in the 1960s and 1970s, it has become a source of concern to the UK Government in terms of its likely impact on airport capacity requirements, land utilisation, social effects and particularly its environmental effects. It is with both the social and environmental issues implied by the UK Government forecasts and by the latter's potential response to these that this paper is primarily concerned.

There is no doubt, at least in the environmental sphere, that the political context has been and continues to be an extremely strong determinant of the *raison d'etre* of much Government-sponsored research in the UK. It could be argued that since 1997 the new Government has fully embraced the dire warnings of global warming in relation to almost all forms of mechanised transport.

However, there is also an element of contradiction in the case of air transport where the UK has consistently been arguing for the adoption of an open skies policy in the EU to foster competition. This is hardly consistent with the threat of demand constraining policies!

It is within this potentially contradictory context that current policy on UK air travel is being formulated. Hence, it is very important to consider a number of aspects of the 'problem' as it has been perceived by the Government. This is because the forecasts for the next twenty years may be sufficiently in error such that any further costs imposed on the industry and or the passengers (as a result of the forecasts) may be significantly out of proportion to the 'problem'. Such an outcome will damage both the industry in the UK and the UK's competitiveness relative to other developed economies. In Scotland we have seen significant growth in both passenger numbers and freight traffic in the 1990s and there is no reason to expect this to wane in the absence of policy intervention. This is especially the case if the forecasts understate passenger growth since restrictive policies are likely to have an even larger impact on UK competitiveness than anticipated. In this paper, we examine a restricted set of questions in relation to the current forecasts of air passenger traffic to and from the UK. These are:

1. How accurate have past forecasts been?
2. How accurate are current forecasts likely to be?
3. Do alternative forecasting methods produce different results?

4. Should future demand be constrained?
5. What are the implications for social equity and regional competitiveness?

These questions are addressed separately in the following sections of the paper. First, it is useful to present some descriptive statistics of the trends in air passenger traffic in the UK.

(Version 2)

1. Introduction

The DETR has requested responses to its consultation document on the future of air transport services in the United Kingdom. A number of alternative policy responses are set out, which are rooted in the DETR's most recent forecasts for passenger growth. The forecasts of demand for air travel are and will continue to be at the core of these policies. However it is argued in this paper that potential policy responses to the forecasts run the distinct risk of being contradictory, socially divisive and a threat to the continued expansion of air travel services in Scotland and also in all regions outwith London.

In the case of Scotland the 'demand constraint' option runs counter to the recent argument from the Secretary of State for Scotland that the country is still poorly served in terms of air transport. This article is concerned with setting out the context of the issues, the efficacy of the forecasts upon which the 'policy options' are being mooted and the implications for Scottish air travel if several of these options are taken up. First it is useful to present some background information on UK air passenger growth.

Between 1974 and 1999 the number of air passengers travelling into and out of the United Kingdom (UK) has increased from 49 million, an increase of almost 250 per cent. The average annual increase in passenger numbers has been slowing down since the 1960s (Table 1):

Table 1 Annual Average Growth Rate in Passenger Numbers

Decade	1960–69	1970–79	1980–89	1990–99	2000–2010
Growth Rate %	14	7.3	5.7	5.1	4.3

The declining rate of growth reflects a consistent movement towards market maturity in passenger air travel for the UK. The decline is expected to continue in the present decade towards an annual average growth rate of 4.3 per cent and is expected to stabilise around this rate for the foreseeable future. This is consistent with forecasts suggesting a doubling in passenger air traffic over the next ten years and a near doubling in the last ten years on a *global* basis. Although the expected lower rate of growth for the UK in the next ten years is considerably less than that recorded in the 1960s and 1970s it has become a source of concern to the UK Government in terms of its likely impact on airport capacity requirements, land utilisation, social effects and particularly its environmental effects. It is with both the social and environmental issues implied by the UK Government forecasts and by the latter's potential response to these that this paper is primarily concerned.

> There is no doubt, at least in the environmental sphere, that the political context has been and continues to be an extremely strong determinant of the *raison d'etre* of much Government-sponsored research in the UK. It could be argued that the new Government has fully embraced the dire warnings of global warming in relation to almost all forms of mechanised transport.
>
> However, there is also an element of contradiction in the case of air transport where the UK has consistently been arguing for the adoption of an open skies policy in the EU to foster competition. This is hardly consistent with the threat of demand constraining policies!
>
> It is within this potentially contradictory context that current policy on UK air travel is being formulated. Hence, it is very important to consider a number of aspects of the 'problem' as it has been perceived by the Government. This is because the forecasts for the next twenty years may be sufficiently in error such that any further costs imposed on the industry and or the passengers (as a result of the forecasts) may be significantly out of proportion to the 'problem'. Such an outcome will damage both the industry in the UK and the UK's competitiveness relative to other developed economies. In Scotland we have seen significant growth in both passenger numbers and freight traffic in the 1990s and there is no reason to expect this to wane in the absence of policy intervention. This is especially the case if the forecasts understate passenger growth since restrictive policies are likely to have an even larger impact on UK competitiveness than anticipated. In this paper we examine a restricted set of questions in relation to the current forecasts of air passenger traffic to and from the UK.
>
> These questions are addressed separately in the following sections of the paper. First, it is useful to present some descriptive statistics of the trends in air passenger traffic in the UK.

You should now be in a position to begin to plan and implement a literature review. But before doing so for your Masters Dissertation you have an opportunity to undertake a critical reading assignment—many of the skills of which are highly central to undertaking a literature review itself. In the next Chapter, you will study the basis of critical reading, which will also be very useful in the literature review you will need to undertake, first, for your second assignment (Research Plan—which must include a brief literature review) and second, for your dissertation itself.

4.10 CRITICAL EVALUATION

A critical review involves structuring and building a logical and coherent argument. It should flow smoothly from one point to the next, drawing upon evidence, and where possible, present alternative viewpoints. It might also involve evaluating the quality of the evidence presented to support an argument, not simply describing it. In other words, critical reading helps one to assess the quality of other people's work, their limitations and to give positive indication for future research.

An unconnected list of 'who said what' is not a discussion, even where an extensive list of authors' names and dates is given. Students often make the mistake of assuming that by including references and quotations from books and articles they are engaging in a 'discussion'. Likewise, describing the criticisms made by other authors does not demonstrate a 'critical analysis'.

4.11 CRITICAL ANALYSIS

You need to show you have examined the material in a critical manner. You could:

- Look at the value of the evidence presented.
- Address inconsistent or incompatible evidence stemming from research and seek to explain it.
- Weigh the pros and cons of different positions, supporting that side of an argument if the quality of evidence favours it.
- Try to find original links between different sources or different strands of an argument.
- Show originality by presenting new ideas or interpretations based upon your own understanding of the material.

4.12 CRITICAL READING

To a critical reader, any single text provides but one view of the facts (or version of reality)—it is one individual's view of the subject matter. Critical readers, therefore, recognise not only what a text says but also how the text portrays the subject matter. They recognise the various ways in which each and every text is the unique creation of a unique author.

Having recognised what a text says, critical readers reflect on what the text does:

Is it offering examples? Arguing? Appealing for sympathy? Making a contrast to clarify a point?

Critical readers also infer what the text, as a whole, means, based on their analysis. The goals of critical reading are therefore:

- To recognise an author's purpose

- To understand the tone and persuasive elements
- To recognise bias

4.13 CRITICAL THINKING

We think critically when we:

- Reply on reason rather than emotion.
- Require evidence, ignore no known evidence and follow evidence where it leads.
- Are concerned more with finding the best explanation than being right.
- Analyse apparent confusion and ask questions.
- Weigh the influence of motives and bias.
- Recognise our own assumptions, prejudices, biases or points of view.
- Evaluate all reasonable inferences.
- Consider a variety of possible viewpoints/perspectives.
- Remain open to alternative interpretations.
- Accept a new explanation, model or paradigm because it explains the evidence better, is simpler or has fewer inconsistencies or covers more data.
- Accept new priorities in response to a re-evaluation of evidence.
- Do not reject unpopular views out of hand.
- Recognise the relevance and/or merit of alternative assumptions/perspectives.
- Recognise the extent and weight of evidence.

4.14 CRITICAL QUESTIONS

In thinking critically about what you read, it is useful to answer a range of questions to help focus your thoughts. The variety of questions that might help you can be split into four categories.

Summary and Definition Questions

- What is (are) …?
- Who …?
- When …?
- How much …?

- How many ...?
- What is an example of ...?

Analysis Questions

- How ...?
- Why ...?
- What are the reasons for ...?
- What are the functions of ...?
- What is the process of ...?
- What other examples of ...?
- What are the causes/results of ...?
- What is the relationship between ... and ...?
- How does ... apply to ...?
- What is (are) the problem(s) or conflict(s) or issue(s) ...?
- What are possible solutions/resolutions to these problems or conflicts or issues ...?
- What is the main argument or thesis of ...?
- How is this argument developed ...?
- What evidence or proof or support is offered?
- What are other theories or arguments from the authors?

Hypothesis Questions

- If ... occurs, then what happens ...?
- If ... had happened, then what would be different ...?
- What does theory X predict will happen ...?

Evaluation Questions

- Is ... good or bad?
- Is ... correct or incorrect?
- Is ... effective or ineffective?
- Is ... relevant or irrelevant?
- Is ... clear or unclear?
- Is ... logical or illogical?
- Is ... applicable or not applicable?

- Is ... proven or not proven?
- Is ... ethical or unethical?
- What are the advantages or disadvantages of ...?
- What are the pros and cons of ...?
- What is the best solution to the problem/conflict/issue?
- What should or should not happen?
- Do I agree or disagree?
- What is my opinion of ...?
- What is my support for my opinion?

4.15 CRITICAL REVIEWS

Let us now summarise *what* questions you should be asking yourself when you are undertaking a critical review of a piece of published research. You should use the following as a template for your own review:

- Has the author of the article clearly defined the *research problem*?
- Has the author clearly explained the *purpose* of the research?
- Is a review of *relevant literature* included in the article?
- Is this review *comprehensive* or too brief?
- Has the author presented and explained a *theoretical framework*?
- Is the *research method*(s) clearly described and explained?
- Are the *research questions* OR *hypotheses* clearly spelled out?
- Is the *data analysis method* clearly explained?
- Is it consistent with the *type of data* (if any) being presented?
- Is the analysis of the data nearly *descriptive* or *analytical*?
- Are the *results* explained in terms of the original *research questions* OR *hypotheses*?
- Are pointers to *further research* given?
- Do the *conclusions* make sense in terms of the *purpose* of the research?
- Does the author spell out the *limits* of the research?
- Is a comprehensive *reference list* given?

Think back to the section in Chapter 2 on research methodology—many of the philosophical and theoretical issues raised there lead directly to the questions above. In undertaking a critical review of a single piece of work or in undertaking a literature review for a dissertation the questions above, if approached in a logical and structured manner, will enable you to work your way through any published research in an efficient and disciplined context.

4.16 EXERCISE

Exercise 1

Select a Journal article (any Journal related to social science or business) and use the template above to critically review it. Discuss the article and your findings in class.

CHAPTER 5

Sampling

5.1 INTRODUCTION

Research design is the blueprint for fulfilling research objectives and answering research questions. In other words, it is a master plan specifying the methods and procedures for collecting and analysing the needed information. In addition, it must ensure that the information collected is appropriate for solving a problem. Therefore, the researcher must have a clear knowledge about the sources of information, the design technique such as survey or experiment, the sampling methodology and the schedule, as well as the cost involved.

The choice of research strategy, that is, *how* a piece of research will be carried out in practice, is fundamentally related to the nature of the research question(s) asked. For example, we may simply be interested in what proportion of the 20–30-year-old age group attends the local cinema but are not interested in *why*—in this case, we require a simple descriptive research design which will be based upon collecting a very narrow set of data. If we are also interested in *why* they attend the local cinema, then we require, at the very least, an exploratory research design which will include detailed surveys and analyses. Therefore, the nature of the research questions largely determines how the research itself will be implemented. The degree to which we are able to answer these questions will further depend on the depth of the research. That is to say, there is a difference between simply describing a phenomenon and explaining it. In order to explain it we need a very deep understanding of the interaction between variables, which constitute the basis of peoples' behaviour. In practice, this is seldom achieved, which is why we need to be very aware of the Limits of Validity of any research design. What do we mean by the 'limits of validity'? In Chapter 2 we discussed the problem syllogistic reasoning where what appears to be a logical argument is in fact not so. In addition, even if we have a logical argument or statement it does not follow that it will be materially correct.

The limits of validity relate to the consistency between logical and material truth and to the rules of evidence as discussed in Chapter 2. In economics, for example, we are told that demand for goods is dependent on price (and a few other things, but mainly price). Hence, if price rises, the demand for most goods will fall and if price falls, the demand will rise. However, this imposes no limits of validity on the demand function. In order for the previous statement to be both logically and materially true, we require that some restrictions be imposed. Hence, price cannot be negative (less than zero) and there must be a maximum price for all goods, which will 'choke' off demand completely—meaning no one is prepared to pay that price. Hence, the limits of validity for any demand function must be as discussed below.

Demand is a function of Price subject to: ($0 \leq P \leq P$ max at Demand = 0). All this says is that price cannot be less than zero and it cannot be greater than the price that chokes off demand completely. If you ever studied physics at school, you will remember that the quadratic function will produce two answers—a negative one and a positive one. If you were measuring speed for example you know, logically, that speed cannot be negative because in the real world negative speed does not exist. Hence, you are intuitively imposing limits of validity on the answers. To put this in very simple language it basically means that when you get an answer to a quantitative or a qualitative question and it instantly 'feels' wrong or just silly then it probably is! So, in terms of research design the key issues facing the researcher are those of validity, reliability and generalisability.

The type of research you undertake will, therefore, have important implications on how you gather and analyse information. The validity, reliability and generalisability of your study will all be influenced by the quality of the information gathered and the methods used to gather this information. This is a critical aspect of the whole research process, and failure to address this issue correctly can have serious consequences for any findings generated from your work.

5.2 CLASSIFICATION OF RESEARCH DESIGNS

Research designs are classified into three groups, as discussed below.

Research Designs in Terms of the Controlling Method

Depending on the controlling method of the research design, research designs are categorised into three types: (i) Experimental design, (ii) Quasi-experimental design and (iii) Non-experimental or observational design.

Experimental Design

In experimental design, researchers plan to measure the response variable depending on the explanatory variable. The response variable is an outcome measure for predicting or forecasting purposes of a study. It is also called dependent variable or predicted variable. Any variable that explains the response variable is called explanatory variable. It is also called independent variable or predictor variable.

The most important factor in the experimental design is randomisation. Clinical trial may be the best example of experimental design, which is nowadays popular in medical statistics.

Quasi-experimental Design

Although the researcher plans to measure the response variable depending on the explanatory variable, there is a lack of randomisation in the quasi-experimental design. It is a mixed design where random and non-random experiments are employed together. For example, fluoride may be found in tap water in certain geographic areas but not in others. The effects on dental decay can be investigated using many methods and a common approach here is to use a quasi-experimental design.

Observational Design

The observational study may be either of prospective or of retrospective design. If the researcher begins to observe, waiting for the results, it is a prospective design. If he/she gathers data at one time and traces the differences into the past, it is a retrospective design (see Figure 5.1).

Research Designs in Terms of Time Sequences

Research designs in terms of time sequences include (i) Prospective design and (ii) Retrospective design.

Prospective Design

In a prospective design, the researcher follows the participants and measures or observes the behaviour of the participants. Depending on the use of randomisation, the

Figure 5.1 Relationships among Research Designs

```
                        Research Design
           ┌───────────────┬───────────────┐
   Experimental      Quasi-experimental   Observational
      Design              Design              Design
        │                   │                   │
   Prospective         Prospective         Retrospective
     Design              Design               Design
        │                   │            ┌──────┴──────┐
   Clinical trial      Cohort design   Case-control   Cross-sectional
                                                          design
```

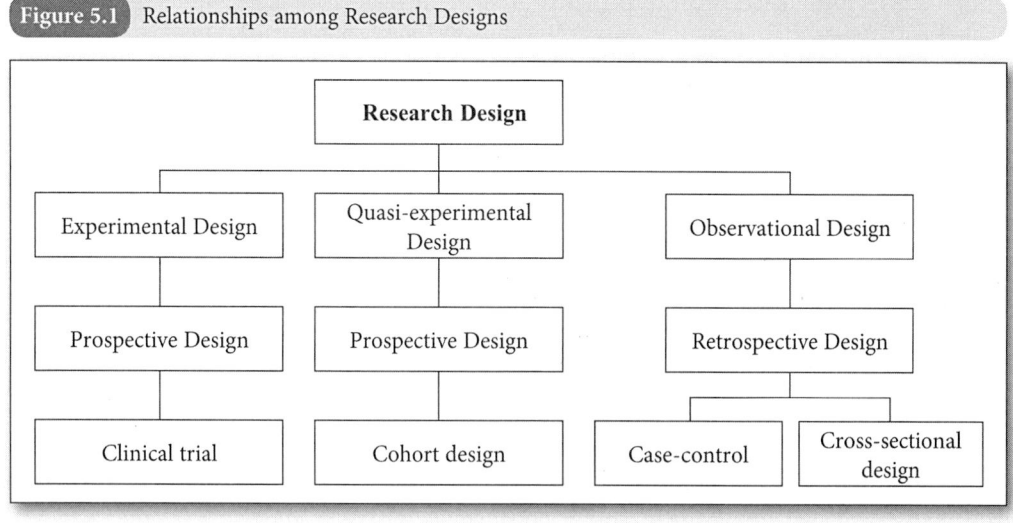

SOURCE: Authors' own.

prospective design is categorised into clinical trials or cohort design. The researcher awaits future events in both designs.

Retrospective Design

In a retrospective design, the researcher gathers data at once and classifies the participants simultaneously into the group categories. If there are only two categories such as 'yes' (case) and 'no' (control) group, these are called case-control studies. If there are more than two categories, then these are called cross-sectional studies.

Research Designs in Terms of Sampling Methods

Research designs in terms of sampling methods include (i) Clinical trial; (ii) Cohort study; (iii) Case-control study and (iv) Cross-sectional study.

Clinical Trial

In clinical trials, the researcher randomly allocates participants to the various groups of interest and measures differences in the future. For example, the researcher randomly assigns 100 students to two different maths programmes. At the end of the term, students are counted according to their outcome of results: pass or fail. These are displayed in Table 5.1.

The odds ratio is (72 * 39)/(61 * 28) = 1.64, which means that the odds of passing in the new programme is 1.64 times the odds of passing in the old programme. It indicates that new programme is more effective than the old one.

Cohort Study

In cohort studies, there is no random assignment. The participants have a right to choose the group they want to join. The researcher measures differences between groups without randomisation in the cohort design. For example, the researcher wants to measure the effect of folic acid on reducing the risk of further strokes among people who have already suffered a stroke. Then volunteers are selected comprising those who are willing to take folic acid. During a follow-up period averaging a couple of years, the numbers of deaths due to heart-related disease are counted to measure the effect of folic acid on stroke. The results are shown in Table 5.2.

The odds ratio is (11 * 685)/(37 * 670) = 0.30, which means that the odds of death in the folic acid group is 0.30 times the odds of death in the vitamin group. This indicates that taking folic acid seems to reduce the chance of having a stroke again.

Case-control Study

In case-control studies, the researcher gathers the data at once and then looks into the past of the participants to classify them, for example, to study whether smoking is related

Table 5.1 Outcomes of Trial

	Pass	Fail
New maths programme	72	28
Old maths programme	61	39

SOURCE: Authors' own.

Table 5.2 Folic Acid Trial Results

	Death	Survive
Folic acid group	11	670
Vitamin group	37	685

SOURCE: Authors' own.

to lung cancer or not. The researcher gathers data at once and then classifies the participants simultaneously into four categories (smoker-cancer group; non-smoker-cancer group; smoker-no cancer group; and non-smoker-no cancer group) (see Table 5.3).

The odds ratio is 3.28, which indicates that the odds of having lung cancer in the smoking group is 3.28 times the odds of having lung cancer in the non-smoking group. The study suggests, therefore, that smoking causes lung cancer.

Cross-sectional Study

The researcher gathers the data at once like case-control studies and then classifies them simultaneously on the classification (more than two categories) and their current responses. For example, the researcher wants to know the relationship between mathematics performance and the number of hours spent watching television a day. He/she can gather information on the average number of hours spent watching television (TV) and the performance in the mathematics exam. As the data are collected at a particular time period and students classified on the basis of the number of hours as well as by maths performance, so it comes under the cross-sectional design. The results are displayed in Table 5.4.

In this case, four odds can be calculated, and if we consider group 1 as a baseline group then we can have the odds ratios of four groups relative to the baseline group. For example, the odds ratio of group 1 and group 2 is $(n_{21} * n_{12})/(n_{11} * n_{22})$. Likewise, we can

Table 5.3 Smoking Study Results

Smoking	Case (Lung Cancer)	Control (No Cancer)
Yes	200	205
No	110	370

SOURCE: Authors' own.

Table 5.4 Cross-sectional Design

Time	Pass	Fail
<1 hour a day (group 1)	n_{11}	n_{12}
2 hours a day (group 2)	n_{21}	n_{22}
3 hours a day (group 3)	n_{31}	n_{32}
4 hours a day (group 4)	n_{41}	n_{42}
5+ hours a day (group 5)	n_{51}	n_{52}

SOURCE: Authors' own.

calculate the odds ratio for the others. The result will allow us to make a cross comparison about students' maths performance and hours spent watching TV.

5.3 SOURCES OF DATA

Data are the facts and figures collected for records or any statistical investigation. Variable is the characteristic of study, which may vary, for example, age, gender, height, employment status, income and so forth. There are primarily two sources of information normally used for research purposes—primary and secondary sources of data. Primary sources are those in which we need to conduct a new survey for gathering information at different levels with regard to the inquiry. Secondary sources are those which are made available or have been collected for other research purposes. Within secondary data exploration, a researcher should start first with an organisation's own data archives. Reports of prior research studies often reveal an extensive amount of historical data or decision-making patterns. Primary and secondary data sources are discussed in detail in Chapters 6 and 7.

5.4 TYPES OF DATA AND MEASUREMENT

Basically, there are two types of data: qualitative and quantitative. Qualitative data are numerically non-measurable; quantitative data can be measured numerically. Most statistical analysis is based on quantitative data using appropriate measurement of their variables. Quantitative variables are also classified into two types: discrete and continuous. A discrete variable can take only certain distinct or isolated values in a given range, for example, number of siblings 0, 1, 2, …, 10. A continuous variable can take any value in a given range, for example, age from 0 year to 100 years. To take another example, if one would like to know what factors are associated with a sales representative's performance, a number of measures might be used to indicate success. Dollar or unit sales volume, or share of accounts lost could be utilised as measures of a salesperson's success. Principally, to enable ease of understanding, the quantitative variables are usually measured by various scales. A scale may be defined as a measuring tool for appropriate quantification of variables. In other words, a scale is a continuous spectrum or series of categories. Like other research, four types of scales are used in business research. These include nominal, ordinal, interval and ratio scales.

Nominal Scale

A nominal scale is the simplest type of scale. The numbers or letters assigned to objects serve as labels for identification or classification. For example, names, player list, gender are categorical variables; one can put the level 'M' for Male and 'F' for Female, or '1' for male and '2' for female, or '1' for female and '2' for male. Other examples include marital status, religion, race, colour and employment status.

Ordinal Scale

When a nominal scale follows an order then it becomes an ordinal scale. In other words, an ordinal scale arranges objects or categorical variables according to an ordered relationship. Thus, ranking of nominal scales is an essential prior criterion for ordinal scales. A typical ordinal scale in business research asks respondents to rate career opportunities and company brands as 'excellent', 'good', 'fair' or 'poor'. Other examples would be (i) result of examination: first, second, third classes and fail; (ii) quality of products and (iii) social class.

Interval Scale

The interval scale indicates the distance or difference in units between two events. In other words, such scales not only indicate order, but also measure the order or distance in units of equal intervals. It is important to note that the location of the zero point is arbitrary. To take an example, in the price index, the number of the base year is set to be usually 100. Another classic example of an interval scale is the temperature where the initial point is always arbitrary.

Ratio Scale

Ratio scales have absolute rather than relative quantities. In other words, if an interval scale has an absolute zero then it can be classified as a ratio scale. The absolute zero represents a point on the scale where there is an absence of the given attribute. For example, age, money and weights are ratio scales because they possess an absolute zero and interval properties.

5.5 METHODS OF DATA COLLECTION

Once the research design (including the sampling plan) is formalised, the process of collecting information—'survey'—from respondents may begin. In this section the term 'survey' is used to define any research method that seeks to gather information from a group of respondents. For example, you may wish to conduct a survey using a questionnaire, focus groups, interviews by telephone or in person, or any other suitable technique. Obviously, there are many research techniques and there are many methods of data collection.

Often there are two phases to the process of collecting data: pre-testing and the main study. A pre-testing phase, using a small sub-sample, may determine whether the data collection plan for the main study is an appropriate procedure. Thus, a small-scale pre-test study provides an advance opportunity for the investigator to check the data collection form to minimise errors due to improper design elements, such as question wording or sequence. Additional benefits include discovery of confusing interviewing instructions, learning if the questionnaire is too long or too short and uncovering other such field errors. Tabulation of data from the pre-tests provides the researcher with a format of the knowledge that may be gained from the actual study. If the tabulation of the data and statistical tests do not answer the researcher's questions, this may lead the investigator to redesign the study. A detailed discussion of data analysis can be found in later chapters.

A common question in data analysis is, 'what is the unit of analysis?' The researcher has to determine the unit of analysis in relation to his/her research problems. The researcher must specify whether the level of investigation will focus on the collection of data about organisations, departments, work groups, individuals or objects. In studies of home-buying, for example, the husband–wife rather than the individual is typically the unit of analysis because the purchase decision is jointly made by the husband and wife. In studies of organisational behaviour, cross-functional teams rather than individual employees may be selected as the unit of analysis. In sociological studies, if the births are counted at an individual level then women are the unit of analysis; if we are interested in aggregate births per household, then the household is the unit of analysis.

5.6 SAMPLING TECHNIQUES

Sampling is the process or technique of selecting a suitable sample for the purpose of determining parameters or characteristics of the whole population. To carry out a study, one might bear in mind what size the sample should be, whether the size is statistically justified and, lastly, what method of sampling is to be used. As for all sampling, we need to consider the time and cost for the survey, whether it is small-scale or large-scale.

How 'representative' is one's sample may be a common question. Researchers always try to draw a representative sample to draw any conclusion about the 'real world'. This is a part of the researcher's responsibility. There are two basic sampling techniques: probability and non-probability sampling. A probability sample is defined as a sample in which every element of the population has an equal chance of being selected. Alternatively, if sample units are selected on the basis of personal judgement, the sample method is a non-probability sample.

A sampling frame is the list of elements from which the sample may be drawn. A simple example of a sampling frame might be a list of all members of an institute or workers in a company or a particular type of company.

The sampling unit is a single element or group of elements subject to selection in the sample. For example, if an airline wishes to sample passengers, every 20th name on a complete list of passengers may be taken. Alternatively, flights can be selected as sampling units. The term 'primary sampling units' (PSUs) designates units selected in the first stage of sampling. If successive stages of sampling are conducted, sampling units are called secondary sampling units, or tertiary sampling units (if three stages are necessary).

5.7 REPRESENTATIVE SAMPLING PLANS

Simple Random Sample

A random sample is defined as follows:

- Selections are made from a specified and defined population (i.e., the frame is known).
- Each unit is selected with known and non-zero probability, so that every unit in the population has an equal (known) chance of selection.
- The method of selection is specified, objective and replicable.

Random sampling results in the selection of a determinate set of units. Substituting other units for those already selected is not allowed at the sampling or data collection stages. A rate of response can then be calculated by the number of responding units divided by the number of eligible units selected. When samples are drawn from small finite populations, an issue of statistical importance arises—whether a unit, once selected, should remain in the population and be given further chances of selection. Social surveys, however, usually select small samples from large populations, so it makes very little difference whether we sample 'with replacement' or without.

Stratified Random Sampling

Simple random sampling can be applied to homogeneous populations in nature; for example, in a school, students' IQs are assumed to be similar for a particular class section. However, in practice, the population (to take the example of school students) is observed to be heterogeneous in nature. Then, in order to apply simple random techniques to such a heterogeneous population, we have to group them as homogeneously as possible, where each group is termed a 'stratum' (in plural 'strata'). Then samples are drawn equally or proportionately from each stratum and, therefore, the procedure is called stratified random sampling. For example, if there are 10 strata and if the requirement is to collect a sample of size 100, one can then draw 10 samples from each stratum. However, if the size of the stratum varies, it would be appropriate to select samples proportionate to stratum size.

Systematic (Quasi-random) Sampling

In most cases, we use systematic random sampling, which guarantees that units cannot be sampled more than once. In systematic random sampling, we range the population from which selections are to be made in a list or series, choose a random staring point and then count through the list selecting every n-th unit.

Cluster (Multistage) Sampling

In cluster sampling we have to have a number of clusters which are characterised by heterogeneity in between and homogeneity within. Cluster sampling is used for a variety of purposes, particularly for large sample surveys or a nation-wide survey. It is very convenient with respect to the time and money allocated for a particular study. The sample is also reliable as it allows random allocation at different stages. If we consider two stages to conduct the survey, then it is called two-stage cluster sampling. If someone considers more than two stages to collect the data, then it is called multistage sampling.

Sequential (Multiphase) Sampling

This is a sampling scheme where the researcher is allowed to draw sample on more than one occasions. It may be economically more convenient to collect information by a sample and then use this information as a basis for selecting a sub-sample for further study. This procedure is called double sampling, multiphase sampling or sequential sampling.

This is a technique frequently used to draw samples in industries for ensuring the quality of their products.

Non-probability Sampling Methods

In non-probability sampling, the probability of selecting population elements is unknown. However, in a situation when a sampling frame is absent, one can easily go for non-probability sampling methods to serve the objectives of the study. However, a question may arise as to how closely these approximate for representativeness. Additional reasons for choosing non-probability over probability sampling are cost and time factors.

Convenience Sampling

Non-probability samples that are unrestricted are called convenience samples. They are the least reliable design but, normally, the cheapest and easiest to conduct. Interviewers have the sole freedom to choose whomever they find, thus the name convenience. Examples include the opinion of people about public transportation systems and customer satisfaction regarding goods and services.

Purposive Sampling

A non-probability sample that conforms to certain criteria is called purposive sampling. There are two major types within this type of sampling:

- Judgement sampling
- Quota sampling

Judgement Sampling

A cross-section of the sample selected by the researcher conforms to some criteria. For example, for election prediction purposes samples are made from those who have previous experience of making election predictions. Judgement sampling is appropriate at the initial stage of research. When one wishes to select a biased group for screening purposes, this sampling method is also a good choice. For example, companies often try out new product ideas on their employees. The rationale is that one would expect the firm's employees to be more favourably disposed towards a new product idea than the public.

If the product does not pass this group, it does not have the prospect of success in the general market.

Quota Sampling

This is often used to improve representativeness of the study. The logic behind quota sampling is that certain relevant characteristics describe the dimensions of the population. In most quota samples, researchers specify more than one control dimension. Each should meet two tests: (i) to have a distribution in the population that will be estimated; and (ii) to be pertinent to the topic studied. In short, researchers should control characteristics such as gender dimension (male, female), religious affiliation and social status in order to draw a representative sample of the population.

Quota sampling is also widely used in surveys, particularly in the commercial world. The main motive for using quota sampling is to reduce the cost of the surveys and the time required to complete them by using a convenient sample of persons who are ready and willing to be interviewed the first time the interviewer calls. However, quotas are imposed specifying that each interviewer must obtain a particular number of interviews with people in each of a small number of subgroups.

Many quota schemes are relatively complex but they all encounter the problem that, within each subgroup, the sample will be biased against those who are not ready and willing to take part in the survey. Such persons are effectively replaced with more available and amenable persons. To sum up, compared with these other methods, random sampling

- Provides protection against selection bias
- Enables the precision of estimates to be estimated
- Enables the scope for non-response bias to be addressed (sampling methods that replace non-responding units conceal this)

Snowball (Network or Chain) Sampling

This is a special type of non-probability sampling where respondents are difficult to identify (who to ask for and what criteria they should have in terms of possessions: rich, poor, homeless, etc.) and are best located through referral networks. Perhaps, it can only be used when the target sample members are involved in some kind of network with others who share the characteristic of interest. A small number of the samples initially selected by the researcher are then asked to nominate a group who would be prepared to be interviewed for the research; these in turn nominate others, and so forth. Reduced sample sizes and costs are a clear advantage in snowball sampling. However, bias is likely to enter into the study because a person who is known to someone has a higher probability of being similar to the first person.

Variations on snowball sampling have been used to study people engaged in illegal activities such as drug cultures, teenage gang activities, power elite, community relations and other applications where respondents are difficult to identify and contact. It is therefore useful when the potential subjects of the research are likely to be sceptical of the researcher's intentions (Arber 2001).

Design Effect

The estimation of sampling errors for stratified and multistage samples is more complex than for simple random samples. For the same sample size, a stratified random sample provides less standard error than for a simple random sample, whereas the standard error for a clustered sample is observed to be greater than that for a simple random sample. The relationship between the standard error of a complex sample design and that of a simple random sample of the same size is called the 'design effect'. The design effect, therefore, measures the effect of the sample design on the precision of population estimates.

5.8 SAMPLE SIZE DETERMINATION

By now, we are familiar with different methods on how to draw a sample from a specified population. The population may be known or unknown but the researcher has to draw conclusions on the basis of a sample and, therefore, sample size determination is an important element in any survey research, although it is a difficult one. Various questions may arise with regard to this issue. How big should the sample be? How small can we allow it to be? A more relevant issue is how to judge whether the sample size is adequate in relation to the goals of the study. Strictly speaking, exact tests to check whether sample size is adequate for the analysis required can be carried out using statistical methods such as significance tests; however, in many studies, readers who do not have the required statistical skills can use a more common-sense approach to the problem.

Sample size is associated with time and cost. It is on the basis of these two constraints one has to determine a sample, which in turn will be able to produce results that are statistically significant, statistically robust or statistically justified, but, more importantly, representative of the whole population. An undersized sample can be a waste of resources for not having the capability to produce useful results. On the other hand, an oversized sample costs more resources than necessary. The existing literature debates the issue of successful selection and meaningful sample size.

Determining sample size varies for various types of research designs and there are several approaches in practice. For example, one can specify the desired width of a confidence interval and to determine the sample size that achieves that goal, a Bayesian approach can be used where we optimise some utility function—perhaps one that involves both precision of estimation and cost. In practice, one of the most popular approaches to sample-size determination involves studying the power of a test of hypothesis; this is discussed here in the case of cross-sectional studies. The following steps are involved:

- Specify a hypothesis test on a parameter θ (the population mean).
- Specify the level of significance of the test α.
- Specify an *effect size* $\tilde{\theta}$ that reflects an alternative of scientific interest. (This effect size is unknown and hence hypothetical.)
- Specify a target value $\tilde{\pi}$ of the power of the test when $\theta = \tilde{\theta}$.

The power of the test is a function $\pi(\theta, n, \alpha)$ where n is the sample size. The required sample size is the smallest integer n such that $\pi(\theta, n, \alpha) \geq \tilde{\pi}$. Moreover, determination of sample sizes for mean and proportion can be calculated under the normality conditions where standard error plays an important role.

Example 1

Suppose a study is conducted on 200 men about their weekly beer consumption and the following results are obtained:

Arithmetic mean = 5.6 pints per week, standard deviation (SD) = 2.1 pints, sample size = 200.

And hence the standard error of mean = $\dfrac{SD}{\sqrt{Sample\ Size}} = \dfrac{2.1}{\sqrt{200}} = 0.148$ pints

A question may often arise, *what is the practical use of the standard error?* It is, however, a key statistic for assessing the precision of the arithmetic mean of a sample.

5.9 TEST OF SIGNIFICANCE FOR POPULATION MEAN

The formula for determining sample size in the case of testing hypothesis of population means can be expressed as follows:

$$n_0 = \alpha_{\alpha/2}^2 \frac{(SD)^2}{d^2}$$

where n_0 = Sample size
Z = Standardised normal value, usually taken as 1.96 for a 95 per cent confidence interval
α = Level of significance
SD = Standard deviation (assumed to be known from prior survey or can be guessed or other published studies can inform on this)
d = Precision range (the required confidence interval)

Example 2

Assume that a researcher wants the estimate to be within ±£25 of the true population value and he/she wishes to be 95 per cent confident it will contain the true population mean. Also assume that early studies have demonstrated the standard deviation to be around £100. What would be the required sample size?

Solution: We are given $Z = 1.96$, SD = 100, $d = 25$.

Thus, $n_0 = (1.96)^2 \dfrac{(100)^2}{25^2}$

$= 64$

5.10 TEST OF SIGNIFICANCE FOR POPULATION PROPORTION

Often the population proportion (P) is another parameter of interest, for example percentage of voters, percentage with a specific interest, prevalence rates of a disease and so on. An example, if there is a binary response in a survey such as respondents categorised as buyer and non-buyer, then the mean of the sample is the proportion (percentage) who are buyers. The standard error of proportion is as follows:

$$SE = \sqrt{\dfrac{p(1-p)}{n_0}}$$

To estimate sample size, an estimate of the population proportion is also needed. Similarly, a statistically valid sample size may be computed by the following formula:

$$n_0 = Z_{\alpha/2}^2 \dfrac{p(1-p)}{d^2}$$

where n_0 = Sample size
Z = Standardised normal value
α = Level of significance
p = Estimated rate
d = Precision range

Example 3

Let us assume that a political party wants to conduct a poll to estimate the percentage of voting for the party within ±5 per cent points and that the party wishes to be 95 per cent confident of the result. Also assume that the percentage of voting for the party is believed to be 35 per cent. What sample is appropriate in this case?

Solution: We have $Z = 1.96$, $p = 0.35$, $d = 0.05$.
Thus

$$n_0 = (1.96)^2 \frac{0.35(1-0.35)}{(0.05)^2}$$

$$= 350$$

Example 4

Suppose we want to estimate the prevalence of tuberculosis (TB) in a city. Since diagnosing TB implies a bacterial culture, X-rays, questionnaire, etc., the process is very expensive. So, we really want to use as few subjects as possible. What precision do we want in our estimate? Let's say 2 per cent. What is our guess of the prevalence? If we do not know it, it is better to err on the safe side and assume 50 per cent. Determine the sample size.

Solution: Using appropriate formula, we have

$$n_0 = (1.96)^2 \frac{0.5(1-0.5)}{(0.02)^2}$$

$$= 2{,}401$$

In any study if the initial sample size is relatively very small compared with the population size (N), which is assumed to be known, then an adjustment is needed using finite population correction (fpc) in order to obtain the required sample size.

The formula is given by

$$n = \frac{Nn_0}{N+n_0}$$

This calculation is assumed for simple random sample (SRS). However, a modification is needed for other sample designs. This modification consists of multiplying the SRS sample size by the design effect.

Thus, a sample size can be guessed for a particular variable of the study. Sample size plays a vital role in determining the precision of sample survey results. If we wish to double the precision of our sample survey estimates (i.e., considering confidence interval width as half), the ultimate sample size and cost will be increased tremendously.

Example 5

Assume that a survey result obtained is that 50 per cent of people smoke cigarettes. The 95 per cent confidence interval is calculated in the normal way and summary results are as shown in Table 5.5.

It is assumed that an interview costs £20 per head. After doubling the precision we can see the immediate effects on sample size as well as on total costs of the survey. Therefore, it may be concluded that increasing precision is not always feasible for a study (see Table 5.6).

For various sample size calculations, one can visit the following useful websites:

http://stat.ubc.ca/~rollin/stats/ssize/index.html
http://www.surveysystem.com/sscalc.htm

Table 5.5 Smoking Survey with Approximately 1 Per cent Precision

Sample Size	Survey Statistic	95% Confidence Limit	Total Survey Cost
8,000	50%	±1.1	£160,000

SOURCE: Authors' own.

Table 5.6 Smoking Survey Doubling the Precision

Sample Size	Survey Statistic	95% Confidence Limit	Total Survey Cost
31,749	50%	±0.55	£634,980

SOURCE: Authors' own.

It is often the case where following a statistical design is not feasible, the population simply might not be large enough to meet the sample requirements, it is simply too difficult doing random sampling due to costs of accessing the sample or the dynamism of the business environment might mean that the business changes before the sample measurement is completed. Indeed, in many student projects one has to make compromises in relation to samples. However, try to ensure that bias is minimised in relation to population proportions being preserved. For example if one samples 100 students about their degree of satisfaction with their programme of study and this sample comprises 70 males and 30 females one would question the bias in the sample if the population of students at the particular university were 56 per cent female.

Try to make the sample representative of the population. In terms of sample sizes for student projects you may not have the resources for a statistical design, our advice is if you are undertaking quantitative research then a minimum of 30 is needed as this might allow statistical tests to be used. If you wish to compare groups then try to ensure that there are at least 20 in each group. So if you are investigating attitudes of males to females and for two different age groups—those aged 30–40 years old with those aged over 50 years old, you need a minimum of 80 in your sample. However, not following statistical designs compromises validity and might lead to incorrect conclusions.

5.11 KEY STATISTICAL CONCEPTS

Statistical Estimates

It is relatively straightforward to gather information about small subsets of populations (e.g., employees of a particular small- or medium-sized company). Large or geographically dispersed populations present more problems. For example, in order to form and monitor their policies, government departments often need quantitative information about populations such as, 'All adults in London', 'Passengers travelling through Heathrow airport' and 'All those eligible to receive a certain benefit'. The rest of the community needs similar information in order to understand how public money is being spent and what is happening in our society. Users need different types of statistical estimates applicable to the population.

Population or Universe

The population consists of any well-defined set of elements. The most important point about a population is that in principle it can be enumerated (all the members can be listed). This list is called a *sampling frame*. Therefore, a sampling frame is a list of members of the population under investigation and is used to select the sample (a part of the population or universe of enquiry). This list should be as complete as possible. For any survey interview, it is essential to have the sampling frame beforehand. The characteristics of a population such as its population mean (μ) and population standard deviation (σ) are called *population parameters*. The characteristics of the sample such as sample mean and sample variance are called *sample statistics*.

Population Totals

Sometimes the users need an estimate of a population total, for example:

- The total amount of money spent by companies on advertising
- The total number of adults who have some attribute (e.g., being disabled)

Population Means

The population mean or average value of variables is an important characteristic of populations. For example:

- The mean amount of money per week spent by households on food
- The mean number of hours worked per week by public service employees
- The mean diastolic blood pressure for adults

Proportions of the Population Having an Attribute

The statistics most commonly presented in social survey reports are percentages, such as the percentage of respondents who have given a particular answer to a particular question. Percentage tables presented in reports are arrays of such estimates. Percentages are actually a convenient convention for presenting *rates* or *ratios*, obtained by dividing one number by another number (the 'base for percentages') and then multiplying it by 100.

Other Rates and Ratios

Other types of ratio are often quoted in survey reports, for example:

- The proportion of teenagers who own a mobile phone
- The price to earnings ratio of company shares

Means, Proportions and Ratios for Sub-populations

All these statistics—means, proportions and other ratios—can be calculated for the total sample and for sub-samples. Much survey analysis is concerned with displaying and comparing values obtained from sub-samples. The tables published in survey reports are designed to make it easy to compare different sub-samples, or samples from the same population at different points in time.

Other Sample-based Estimates

Many survey reports contain estimates other than the familiar means, percentages and ratios, for example, factors such as the correlation coefficient, multiple regression coefficients and factor scores, among others.

Sources of Statistical Estimates

There are three main ways of obtaining estimates of these types.

- Use of population information already collected for administrative purposes
- Conduct a 100 per cent enumeration or census
- Carry out a sample survey

The first two methods have some advantages, but also have some problems.

Administrative Statistics

- May not focus on the population we want to know about
- May not contain all the information we require
- Use definitions for administrative, rather than statistical or research purposes
- Measure interactions with the administrative system and the inadequacies of that system, rather than what is happening in the whole population
- May be incomplete in their coverage of the target population
- May be out of date or inaccurate, particularly if there is no built-in administrative reason for continually updating or correcting them

Censuses

- Tend to be large, costly, slow and unwieldy
- Use simplified, generalised methods of data collection that cannot handle complicated concepts or focus on diverse subgroups
- Rely on cheap and fallible methods of collecting and recording data and, therefore, suffer from response bias (e.g., respondents misunderstand questions) and from recording and processing errors
- Are conducted rarely, so that the results are usually out of date

Why Sample?

Cost: Rather surprisingly, the sample size required to provide estimates of a given level of precision is virtually independent of the population size. Therefore, samples of a modest size (say, 1,000 drawn from a population of 40 million), if correctly selected, can often give useful results at vastly lower cost than a census. Administrative statistics, if available, may be cheaper still and in principle may be free of sampling error. However, they fall short in other ways described earlier.

Relevance and flexibility: Purpose-designed surveys can adopt the definitions and cover the topics that are most relevant to a particular information need. They may also be able to tailor data collection to different subgroups and different circumstances.

Speed: Surveys can often be mounted and completed and deliver results within a period of months or even weeks. Censuses need long lead-times and administrative statistics are produced to inflexible administrative schedules.

Practicality and feasibility: In many situations it is simply not feasible or practical to set up and run a census or to change administrative procedures so as to produce the required statistics. A sample survey which will meet some of the information need is usually feasible and practical.

Higher data quality: Sample surveys deal with much smaller numbers of units than administrative statistics or censuses. They can, therefore, afford to devote more attention to controlling the systems upon which data quality depends.

Public acceptability: Both administrative statistics and censuses involve bureaucracy and compulsion, which the public dislikes. Voluntary surveys using relatively small samples tend to be more acceptable.

Variance and bias: Sample-based methods of obtaining estimates of these kinds also have some inherent problems. However, there are methods for dealing with these problems and producing the best possible sample-based estimates for a given survey. The results of sample surveys are not completely precise, but suffer from *random sampling variability*. This means that the same sampling method applied repeatedly to the same population will not produce identical results each time. In general, the results of any one sample survey will differ from the true population figures. *Bias* is different from random variation. A method that is unbiased will produce results which vary randomly from sample to sample but which, on average, correctly reflect the population. A biased method will systematically misrepresent the population, no matter how large the sample. The main sources of bias are discussed below.

Imperfect coverage: Like other methods of obtaining information about large dispersed populations, sample surveys may suffer from coverage shortfalls because the source of the sample did not cover the whole of the target population. For example, most household surveys omit persons and households with no fixed address, because of the difficulty of sampling them. Particular types of coverage shortfall may matter a great deal or not at all, according to the purpose of the survey and the size of the shortfall.

Sampling bias: If the sample design or sampling procedure gives certain types of units a higher chance of selection than others, the results will be biased. Subjective judgement in the selection of units is likely to cause bias; however, bias may be present even if the method is objective.

Non-response bias: Voluntary sample surveys conducted by competent survey organisations typically obtain usable responses from proportions of the selected sample that will vary between 60 per cent and 90 per cent. The main reasons for non-response are non-contact and non-cooperation. There is usually a suspicion that sample units which are not contacted, or decline to cooperate, differ from those who do provide information in ways which are relevant to the purpose of the survey. For this reason, keeping response rates up is a major preoccupation of organisations that carry out surveys based on random sampling.

Response and other data collection and processing biases: A strong point of sample surveys over censuses and administrative statistics is that they can address complicated or difficult topics, and can use sophisticated methods of data collection. However, these carry risks that they will themselves introduce biases. This can happen, for example, because respondents misunderstand or do not carry out in a consistent manner the tasks of retrieving and reporting information which the survey imposes upon them (response bias); or because other people involved in handling the information, such as interviewers or coders, introduce biases of their own.

5.12 SOME PROBLEMS WITH RANDOM SAMPLE SURVEYS

Non-response Bias

Non-response bias can distort selection probabilities, so that a sample that started off as 'equal probability' can end up with some sample members being under-represented, while others are over-represented. However, with time and hard efforts, non-response bias can be kept to a certain tolerable level. In addition, there are statistical methods that can (partially) compensate for non-response bias.

Limited Sample Size and the 'Inverse Square Law'

A larger sample size gives more precise estimates; however, precision does not increase directly with sample size, it does so as the square root of the sample size. Therefore, for example, to halve the sampling error you need a sample four times large. In practice, this limits the degree of precision we can achieve. However, we can make estimates somewhat more precise by good sample design, including pre-stratification and post-stratification. In addition, it turns out that many estimates are useful even if not very precise.

Different Estimates Have Different Margins of Sampling Error

Each of the many estimates typically produced by a survey based on a single sample may, in principle, have *different* margins of sampling error. Sampling error is determined not only by sample size, but also by the way in which the particular variable is distributed in the population. However, we can calculate what these margins (or confidence intervals) are, even for complex multistage sample designs.

Sampling Distribution, Sampling Bias and Sampling Variance

Sampling Distribution

A key concept is that of the sampling distribution. Suppose, we are trying to measure the mean of a continuous variable, such as mean income per adult in India (and that we have already developed both conceptual and operational definitions of 'income' and 'adult'). If we decide upon a sample design (and sample size), implement it and carry out a survey, we get an estimate. If we could then carry out the survey again using exactly the same sample design, we might get a different estimate (because the sample will contain a different set of people with different incomes). If we carried on repeating the survey over and over again, we would get a large set of different estimates. The complete set of estimates that could be obtained from all the samples that could be selected under the sample design is known as the *sampling distribution*.

The sampling distribution is simply a statement of what values the estimate can take and the frequency with which it will take them. Given that you are only doing the survey once, the frequency distribution is equivalent to the *probability* that it will take each value. A sampling distribution could be drawn as a histogram or graph, or presented as a table.

Sampling Bias

The intuitive idea of sampling distribution is based on the assumption of equal selection probabilities. Strictly, a sampling distribution is a distribution of the *estimates* obtained by an estimation method (which encompasses the sampling method). If selection probabilities are not equal, it is essential that estimates should be based on *weighted* data where each sample member is weighted by the inverse of their selection probability. Only then is the estimate unbiased.

However, what exactly is meant by bias? If the average of all the estimates that could be obtained, across all possible samples under a particular sample design, equals the actual population parameter being estimated, the design is said to be unbiased. That is, an unbiased design (estimation method) is one which is centred on the actual population value (remember that sampling distributions are symmetrical).

If a sample design is unbiased, that does not imply that the *one* estimate produced by a survey using that design will equal the population value. It simply means that *on average* a sample drawn under the design will produce an estimate equal to the population value.

Therefore, bias is a measure of the *location* of the sampling distribution, relative to the population value; variance is a measure of the *spread* of the distribution.

Sampling Variance

Researchers can influence the sampling distribution, because they can choose the sample design. Ideally, we want the values of the estimates in the sampling distribution to vary as little as possible around the true population value: we want to minimise the chances of our survey producing an estimate that is very different from the true value. This variation around the true value is what is meant by *sampling variance*. The larger the sampling variance, the greater the chance that the survey will come up with an estimate that is different from the population value.

You can calculate the variance of *any* set of numbers. But if that set of numbers happens to be a sampling distribution, then the variance is known as the *sampling variance*, and the square root of the variance is known as the *sampling error* (*standard error*).

Sampling Error

Specifically, sampling error is the amount by which an estimate differs from the population value (due to sampling). The term 'sampling error' is also used in a general sense to refer to the *tendency* for estimates to differ from the population value.

Error can arise at the sampling stage from one of two sources: variance or bias. Hence, *sampling error* is the sum of the effects of bias and variance (but *survey error* is not just sampling error; there are also other sources of error, such as measurement error). Sampling error is a measure of the expected difference between the estimate and the true value. To minimise this expected difference is to maximise the *accuracy* of the estimate.

The sampling distributions arising from different sample designs can be compared in order to determine the design that will produce the most accurate estimate.

If a survey estimate is known to be 'in error', it cannot be determined, from that knowledge alone, whether or not the sample design was biased. Owing to sampling variance, an unbiased sample design could produce an estimate that is in error, or a biased design could produce a 'spot-on' estimate. This is one of the central paradoxes of sampling: it is impossible to know, from examination of the sample alone, whether it is free from selection bias. It is not enough simply to have not *detected* any bias—we need to *ensure* that there is no possibility of arising bias. Conversely, there are no grounds for rejecting a sample, provided that you have confidence in the selection process. In choosing the best sample design, a problem is that most surveys are trying to measure a large number of things—each of which will have a different sampling distribution. So, hard decisions must be made about the relative importance of different estimates. An optimal sample design for one estimate will not necessarily be the best design for some other estimate. As discussed earlier and in Chapter 2, the key issues to be borne in mind in *any* research design concern validity, reliability and generalisability. These are discussed in Chapter 14.

Confidence Intervals

If, for a particular sample design, you know the sampling distribution of the estimate in which you are interested in, then you can see what proportion of sample produces an estimate within plus or minus any given value of the population value. For example, suppose there are only 20 possible samples, and 19 of them would give an estimate (mean income amongst the sample) within HK$100 of the population mean income. Then you might say you were '95 per cent confident' (19/20) that any one sample will produce an estimate within plus or minus HK$100 of the true value. We can loosely think of this as a *confidence interval*. The idea of confidence intervals stems from the idea of the sampling distribution. Statements can be made about the precision of survey estimates. This is often done in the form of n standard errors (commonly $n = 2$), and this is commonly referred to as a confidence interval. But to do this, it is necessary to estimate what the sampling distribution looks like.

5.13 THE NORMAL DISTRIBUTION

The sampling distribution of many statistics, such as proportions and means, has the same shape (provided that the sample size is not too small). This shape is called the *normal distribution*. The normal distribution is a statistical distribution that has a particular shape and known properties. The most important of these properties are that it is symmetrical and that there is a known and fixed relationship between the standard deviation of the distribution (the standard error) and the percentiles of the distribution. In other words, a certain distance from the centre of the distribution, in terms of standard deviations, always represents coverage of the same proportion of the area under the curve.

For example, to proceed one standard deviation from the midpoint will cover 34.1 per cent of the area under the curve. In other words, 68.2 per cent of samples will produce an estimate that is within ±1 standard deviation. Hence, for the one sample that you have drawn, there is a 68.2 per cent chance that the sample statistic is within 1 standard deviation of the population parameter. This is a confidence interval. We are always (for almost any estimate and any unbiased sample design) 68 per cent *confident* that the estimate will be within ±1 standard deviation of the population parameter; 95 per cent confident that the estimate will be within ±2 standard deviations, and so on (the number of standard deviations corresponding to any prescribed level of confidence can be looked up in standard statistical tables). A graph representing the density function of the normal probability distribution is also known as the normal curve (see Figure 5.2). To draw such a curve, one needs to specify two parameters: the mean and standard distribution. A normal distribution with a mean of zero and a standard deviation of 1, that is, ($\mu = 0$, $\sigma = 1$) is also known as standard normal distribution.

It is useful to remember that approximately two thirds of the data lies within plus or minus one standard error of the mean (zero in Figure 5.2). Within plus or minus three standard errors of the mean, 99.7 per cent of the data would be expected. Thus if a data point is further away than three standard errors from the mean, then it would be deemed unusual—an outlier or from another distribution. This forms the basis of many rules in statistical quality improvement.

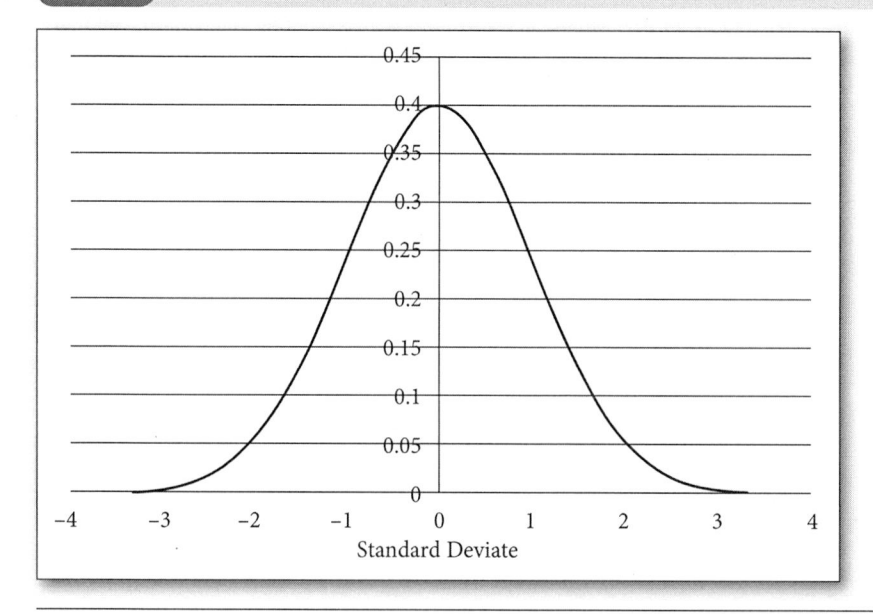

Figure 5.2 The Standardised Normal Distribution

SOURCE: Authors' own.

5.14 EXERCISES

Exercise 1

Provide an example of a cross-sectional study that you are familiar with. How do you think the researchers attempt to make the study representative?

Exercise 2

Give an example of a longitudinal study that you would wish to carry out. What particular problems might you encounter?

5.15 REFERENCE

S. Arber, 'Designing Samples', in *Researching Social Life*, ed. N. Gilbert (London: SAGE Publications, 2001), 58–82.

CHAPTER 6

Primary Data Collection

6.1 INTRODUCTION

A number of approaches to gathering your own original data are outlined in this chapter. This is a very important aspect of research design and the ability to achieve the research aims and answer the research questions depends on the effectiveness of data collection. In student work, you must think about the practicality of obtaining the required data in the available time period and also the accessibility to the field site. This means that careful consideration and planning of data collection are required.

First ask: do I really need to collect data myself or is the data I need to answer the research questions available elsewhere (this is termed secondary data). If it is then use secondary sources if you can get access to it. Often secondary data is easier to use and tends to be more comprehensive, reliable and valid than data that you will be collecting yourself. Using secondary data is the subject of the next chapter, but often it does not answer the particular questions you are pursuing. This will mean you need to collect data yourself, which is the primary data. Collecting primary data is expensive, time-consuming and difficult. In this chapter we outline the main approaches used in business and management research and expand on some of them in subsequent chapters.

6.2 OBSERVATION

Although observation is a data collection method in its own right, no matter which data collection method you follow, observation should be an important element. Some of the most important findings in research have been accidental and captured from observations of the failures of other data collection methods. Thus, be alert and observe and note and document these observations. In order to do this, maintain a research diary.

This should be a small notebook which accompanies you everywhere, and observations, chance findings and important references can be noted in it. In some situations in the field, for instance perhaps while conducting interviews in a company, you might want to make observations about the physical infrastructure, symbolic images of the company and how the staff in the company looks and behave. In marketing you may wish to observe the behaviour of shoppers in a shopping mall. What gets them excited? What shops do they linger at? To make observations more efficient, it is often wise to construct data collection sheets. An example of one is shown in Figure 6.1, which has been used to aid in the observation of people's behaviour in bars and restaurants.

Similarly, in a human resource study one might be interested in participation details in weekly meetings. Therefore, a map of the meeting room could be produced like the one shown in Figure 6.2.

Every week as an observer you can note where people sit, when do they arrive and leave and the degree to which they participate in the meeting. Over the weeks, inferences may be drawn—perhaps one manager is keen to win the boss' attention, and so always sits near him/her and arrives early. Whereas another may be disenfranchised, and so always sits near the door to leave early. Another may not be interested and often daydreams, always getting the space next to the window.

To record what attracts customers in a shopping mall—or to identify shops that they avoid—again a map can be used, an example of which is shown in Figure 6.3.

In conducting observations there is a need to be unobtrusive so that people do not change their behaviour because they are being watched. An important data collection tool

Figure 6.1 Data Sheet to Record Activities in Bars and Restaurants

Site		Date:		
Observer		Start time		End time
		Bar/Restaurant (zone)		
Activity	A	B	C	D
Drink—types and amount				
Food—type and amount				
Chat—to whom				
Chat how long				
Other activity				

SOURCE: Authors' own.

Figure 6.2 Map of Meeting Room

SOURCE: Authors' own.

Figure 6.3 Map to Record How People Shop

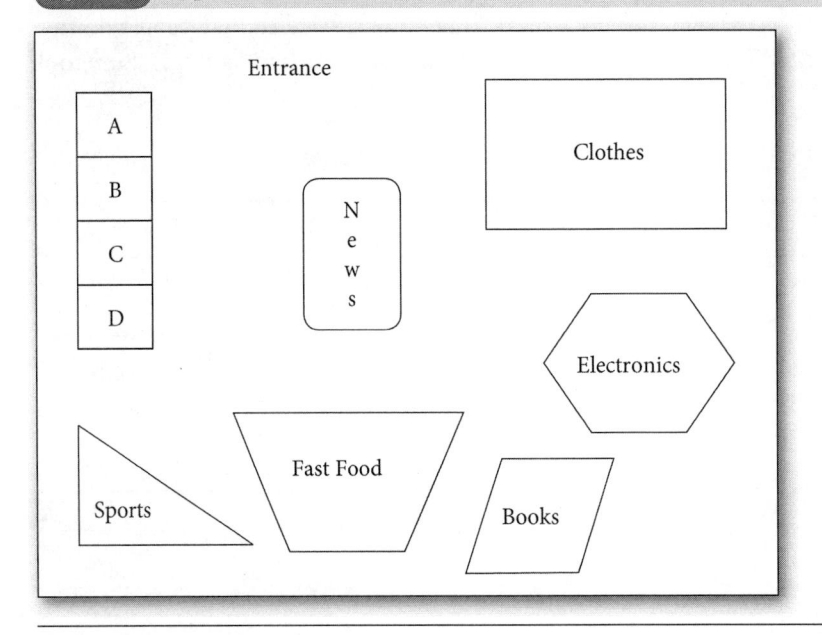

SOURCE: Authors' own.

for research in the workplace is to become part of the organisation and observe the behaviour of colleagues. This is called *participant observation* and the method has its origins in sociology and anthropology and in business. Management research is appropriately applied to the organisation as a 'tribe' where often the intention is to study behaviour. The role of participant observer is not simple and four main types of roles exist:

- Researcher as employee working within the organisation
- Researcher in explicit role of researcher within the organisation
- Interrupted involvement (sporadic involvement over a period of time)
- Observation alone (avoids interaction with the subjects of the research but is limited in the sense that you can only get a description of what takes place but not why)

Overall, there are some serious ethical issues to be dealt with in this method and you should get the permission of those you observe. This may, however, cause them to change their behaviour.

6.3 EXPERIMENTATION

Experimentation is the main tool in the physical researcher's armoury where it has been very successful. The idea is to determine the effects of various factors on a response variable by varying these factors in a controlled way, and often in controlled conditions. Experimentation can be a very reliable and efficient means of collecting data and verifying or refuting theories. Yet, despite the success of experimentation, students make little use of it as a data collection method. Perhaps the reason for this is that statisticians often portray the design of experiments as a very complex procedure. It can be, but for most purposes, simple designs are good enough.

Example: Consider how exam scores are affected by whether students attended lectures or worked from a book, the gender of the student, and if the student had a job. Then one could use the following scheme as shown in Table 6.1.

Here all combinations of levels within the factors are listed, and this is called a *full factorial experiment*. The idea is to assign students to each of the factor level combinations according to the category they fit into. The more the number of students in each category, the more reliable, usually, are the results. In this example, the average response from lecturing is (70 + 60 + 50 + 80)/4 = 65, *while* that of book-based study is 55. Therefore, lecturing appears to be the best approach. Can you ascertain if male students are better than female students or vice versa? What about the effect of working? Much of the statistical literature is concerned with assessing the significance of results.

Table 6.1 Influence of Various Factors on Exam Scores

Factor			
Study Style	Gender	Working	Mean Response
Lecture	Male	No job	70
Lecture	Female	Job	60
Lecture	Male	Job	50
Lecture	Female	No job	80
Book	Male	No job	60
Book	Female	Job	50
Book	Male	Job	50
Book	Female	No job	60

SOURCE: Authors' own.

However, it is the practical, almost common-sense method of being able to control what is important.

Experiments are rarely used in business and management research, perhaps because of the difficulty in controlling the influential factors while the experiment is being conducted. As a consequence, experimentation will not be pursued in this book. Interested readers should consult Box et al. (1978) and Montgomery (2005) for a more technical account.

6.4 SURVEYS

To obtain information from people, it seems obvious that one should either question them face-to-face, or conduct telephonic surveys or mail questionnaires. Indeed, surveys are, perhaps, the most widely used method of data collection in business and management research. As surveys dominate thought, most researchers are reluctant to think of other methods. We encourage you to consider other methods because response rates from surveys—whether postal, telephonic or electronic—are rarely higher than 20 per cent—hardly representative of a population. In conducting surveys the construction and design of the instrument, or the questionnaire, are critically important, as are sample selection and administration. As surveys are so important, we have dedicated a separate chapter to this data collection method (see Chapter 8).

6.5 INTERVIEWS

For many, qualitative data is required to understand in-depth motivations for people's behaviour or feelings. Face-to-face or telephone interviews are frequently used in business and management research. It allows a mass of information to be collected but is very time-consuming and sample sizes tend to be small. Thus, although one obtains in-depth information, one may question the representativeness of the findings. The concept of interviewing can be extended and people interviewed in groups to make the process more efficient. In market research, group interviews are used to focus on a particular aspect and such a data collection method is called *focus groups*. Again, because of the importance of the process of interviewing, a separate chapter is devoted to interviews and focus groups (see Chapter 9).

6.6 DIARY METHODS

Diaries can be either qualitative or quantitative depending on the kind of information that is recorded. At one level they may be a simple record of events from which activity sampling may provide a statistical treatment, while at another level they may take the form of a personal journal research process. Diaries are good to find out about people's consumption, or travel and leisure patterns. Individuals are asked to maintain a diary but it needs a structure to enable the writer to focus on what is relevant. There are several problems in this method, notably those given below:

- There is a need to select people who can express themselves well.
- There is a need to structure to focus the writer.
- Maintaining the diary is time-consuming.
- The writer needs continued encouragement.
- There may be particular anxieties about confidentiality.
- The process requires time to set the flow of writing and to analyse the findings.
- Subjects may frequently forget to enter items and diaries are rarely complete accounts in themselves

For undergraduate or Masters research we do not recommend the use of the diary method, mainly because there is limited study time.

6.7 CASE STUDIES

Case studies are used to study particular phenomena in particular settings. The case study method is very common in business research and is particularly useful for the analysis of organisations. However, it can be narrow in scope and a generalisation can be very difficult. It is often used to determine if a certain approach works in a particular setting. Nevertheless, the case study can be a very powerful research method in terms of questioning accepted theory. This is because it is essentially an inductive research method, which we discussed in Chapter 2. It is rooted in the observation of empirical 'data' and then can be used, within limits, to evaluate the efficacy of particular theoretical frameworks. One difficulty with the method, however, is that ensuring access and continuity to organisations can sometimes be problematic. In case studies, researchers typically use a mixed data collection approach, utilising a combination of observation, surveys and interviews. A good text on the case study approach has been authored by Yin (1994); also read the journal paper by Eisenhardt (1989).

A case study is an in-depth study which explores issues, present and past, as they affect one or more units (organisation, group, department or person). One may use a single case-study design or multiple case studies, which make for a more comparative approach. The comparative case study asks the same questions in several related organisations as well as your own. In a managerial setting, this is commonly referred to as 'benchmarking'. Case studies are often used by those researching operations management who are attempting to identify the 'best practice'. This 'best practice' might be in minimising absenteeism, or implementation of Total Quality Management or constructing and using sales forecasts.

Case studies are to do with uniqueness, understanding and particularisation rather than generalisation. They are naturalistic and field oriented. They ask the questions 'How?' and 'Why?' and the research questions may *evolve* as the research progresses. They *generate* rather than test hypotheses and these hypotheses or models can then be tested possibly by other researchers.

Case studies can include both qualitative and quantitative research. The key advantage is the scope for gathering a rich source of data that allows for particularisation, that is, getting to know the uniqueness of the individual case and its context. Yin (1994) categorises case-study research into the following four main categories:

1. To explain the causal links in real-life interventions that are too complex for the survey or experimental strategies.
2. To describe the real-life context in which an intervention has occurred.
3. An illustrative case study may provide a journalistic account of the intervention.

4. It may be used to explore those situations in which the intervention being evaluated has no clear single set of outcomes.

Purpose of the Case Study

Case studies can be used when it is not possible to adopt a sampling approach that seeks to generalise conclusions as if conditions were identical in other organisations, that is, not possible to generalise from a case study but possible to generalise, or test, a *theory*. For example, you could test the theory that senior manager involvement leads to increased job satisfaction and productivity amongst lower grades of staff. Often the purpose is to *replicate*: to compare the organisation you are studying with others in a systematic way, and to *explore different stances* to the issues you are examining, or *different levels* of the variables involved.

'Small generalisations' can be made about a particular case, for example, generalisations about links between high-involvement practices and organisational performance within a particular organisation context. Case studies may also provide counter examples that invite modification of a grand generalisation.

6.8 DATA STORAGE

Spread sheets, such as Excel, Minitab, SPSS, etc., provide a convenient way for storing data. Typically, each case or respondent is a row of data and each column of data is a variable (or question if using a survey)—this then would form the Data Matrix. Ensure that the files and columns in the spread sheet are given sensible names that will aid your memory. If a data value is missing then just leave the cell blank (the old idea was to code missing or not appropriate answers as 99 or 999; we recommend that this is not done as there is a tendency to forget about the coding and it gets included in the analysis, rather it is far better to leave the cell without a value).

6.9 TRIANGULATION

In business and management there is a need for triangulation in order to search both for accuracy of the data and for alternate explanations. The idea is to collect data by different means and the hope is that there is convergence on the truth. From a qualitative perspective, this process is complex because of the notion of social constructivism (it takes a subjective rather than an objective view of the world). This perspective held by most qualitative researchers rests on the belief that there are *multiple perspectives* or views of the case that need to be represented, and that there is no way to establish, beyond contention, the best view or the 'truth'. In qualitative research, therefore, efforts to find the validity of data observed go beyond simple repetitive action of data gathering. The principles underlying the choice of data collection techniques are based on two key requirements of qualitative research:

- The need to gain full access to the 'knowledge and meanings of informants' (Easterby-Smith et al. 1993)
- To achieve 'plausibility' and 'credibility' of the evidence presented and assertions made by the researcher (Boulton and Hammersley 1996)

Plausibility is defined by Boulton and Hammersley (1996) as "the extent to which a claim seems to be true given its relationship to what we and others take to be knowledge that is beyond reasonable doubt". Credibility is "whether the claim of a kind that, given what we know about how the research was carried out, we can judge it to be very likely to be true". Stake (1995) forwards protocols for triangulation. These are as follows:

Data source triangulation: The analysts ask whether or not what they are reporting is likely to be constant at other times or circumstances.

Investigator triangulation: Other researchers take a look at the same scene. Or findings can be presented to other researchers to discuss alternative interpretations.

Theory triangulation: Multiple investigators agree as to the meaning of the phenomenon.

Methodological triangulation: This involves using a variety of data collection methods to build confidence in the interpretations made so far.

Member triangulation: The respondent is asked to review the material for accuracy and to add further comments that might aid description and explanation. By doing so, the actors personally help triangulate the researcher's observations and interpretations.

These protocols should be considered in your research.

Interpretation as Method	
Quantitative Research	Qualitative Research
• Operationally binds the inquiry to be defined into variables (small number).	• Seeks unanticipated as well as expected relationships.
• Minimises importance of interpretation until data are analysed. Period of data collection and statistical analysis thought of as 'value free'.	• Dependent variables are experientially rather than operationally defined.
	• Even independent variables expected to develop in unexpected ways.
	• Situational conditions are not known in advance.
	• Findings are not so much 'findings' as 'assertions'.

Critique checklist for a case-study report

- Is this report easy to read?
- Does it fit together, each sentence contributing to the whole?
- Does this report have a conceptual structure (i.e., themes or issues)?
- Are its issues developed in a serious and scholarly way?
- Is the case adequately defined?
- Is there a sense of story to the presentation?
- Is the reader provided with some vicarious experience?
- Have quotations been used effectively?
- Are headings, figures, artefacts, appendices, indexes effectively used?
- Was it edited well, then again with a last-minute polish?
- Has the writer made sound assertions, or is there over- or under-interpretation?
- Has adequate attention been paid to various contexts?
- Were sufficient raw data presented?
- Were data sources well chosen and in sufficient number?
- Do observation and interpretations appear to have been triangulated?
- Is the role and point of view of the researcher made apparent?
- Is the nature of the intended audience apparent?
- Is empathy shown for all sides?
- Are personal intentions examined?
- Does it appear individuals were put at risk?

6.10 EXERCISES

Exercise 1

Discuss the strengths and weaknesses of using surveys, case studies and observation to gather primary information.

Exercise 2

The following data relates to the unit cost per journey to the workplace for eight people. From the information provided, determine which of age, gender and travel mode affect the cost of travel, and to what extent.

Factor			
Age	Gender	Travel Mode	Cost per Journey
18	Male	Bus	7
18	Female	Car	9
18	Male	Car	12
18	Female	Bus	9
45	Male	Bus	7
45	Female	Car	13
45	Male	Car	14
45	Female	Bus	9

Exercise 3

- Design a data collection form to allow observation of workers' behaviour in an open-plan office.
- Design a data collection form to allow observation of sales staff interactions with customers.
- Design a data collection form to allow observation of employees in a restaurant.

6.11 REFERENCES

D. Boulton and M. Hammersley, 'Analysis of Unstructured Data', in *Data Collection and Analysis,* ed. R. Sapsford and V. Juup (London: SAGE Publications, 1996), 243–259.

G.E.P. Box, W.G. Hunter and J.S. Hunter, *Statistics for Experimenters: An Introduction to Design, Data Analysis, and Model Building* (Chichester: Wiley, 1978).

M. Easterby-Smith, R. Thorpe and A. Lowe, *Management Research: An Introduction* (London: SAGE Publications, 1993).

K.M. Eisenhardt, 'Building Theories from Case Study Research', *The Academy of Management Review* 14, no. 4 (1989): 532–50. http://pages.cpsc.ucalgary.ca/~sillito/cpsc-601.23/readings/eisenhardt-1989.pdf

D.C. Montgomery, *Design and Analysis of Experiments* (Chichester: Wiley, 2005).

R. Stake, *The Art of Case Research* (Thousand Oaks, CA: SAGE Publications, 1995).

R. Yin, *Case Study Research: Design and Methods* (Second Edition) (Beverly Hills, CA: SAGE Publications, 1994).

6.12 WEBSITES

For more information see:

> http://www.tele.sunyit.edu/traingulation.htm
> http://www.socialresearchmethods.net/tutorial/Brown/lauratp.htm
> http://www.orientpacific.com/observational-techniques.htm
> http://www.nova.edu/ssss/QR/QR3-2/tellis1.html
> http://www.ischool.utexas.edu/~ssoy/usesusers/l391d1b.htm

CHAPTER 7

Secondary Data Collection

7.1 INTRODUCTION

Secondary data is data collected by someone else and there is a great deal available to you from books, libraries and on the web. You can use this data as the main source for your research or as a supplement to the data you collect. Secondary data is often used to validate your sample. For example, you may sample students in a university and find 60 per cent are male; to verify if your sample is representative you could consult the statistics held centrally by the university to compare your sample percentage with the population percentage. If you were studying how the amount of human resources and GDP devoted to health services is associated with the duration of life in different countries, then data from the Human Development Report held at http://hdr.undp.org/statistics/ is possibly sufficient.

The use of secondary sources such as government-produced records, personnel records and financial histories tend to be the quickest (Stewart and Kemins1992). There is an ever-increasing amount of data collections on the web; see for example the following URLs:

http://www.worldbank.org/
http://dir.yahoo.com/Government/
http://www.who.int/
http://www.census.gov/
http://hdr.undp.org/hdr2006/statistics/
http://esa.un.org/unpp/
http://www.esds.ac.uk/search/searchStart.asp
http://www.measuredhs.com/countries/
http://esds.ac.uk

Social Science databases worldwide

http://www.sociosite.net/databases.php#WW

There are some problems with secondary data that have to be considered, such as is it really valid to your work and is it really representative? The data could be on tourist travel last year but you want to predict tourist travel this year. Often secondary data is aggregated to a regional or even a national level, so it is not of much use if you are trying to carry out local comparisons. Sometimes definitions of variables are ambiguous and there are often problems with cohesion over time in that the definition of the variable may have changed several times in the history of that variable. Examples are definitions of unemployment and inflation. The major advantages and disadvantages of secondary data are listed below.

The advantages of using secondary data are:

1. Large representative samples well beyond the resources of the individual researcher are available.
2. Good for examining longitudinal data and looking for trends.
3. Supporting documentation and explanation of methodology, sampling strategy, data codes are given.
4. The researcher can concentrate on data analysis and interpretation.

The disadvantages of secondary data are:

1. *Data compatibility*: does the information match what is required for your research?
2. *Data coverage*: does the information cover all subjects or groups in your research?
3. There can be depth limitation in that you may see a trend or an oddity in a time series but there may be no data available to allow investigation of the reasons or consequences.
4. Does the information come from all time periods or are there gaps?
5. Consistency of time series.
6. Historical and therefore may not be relevant to current issues.
7. Need to assess the quality of the data and the approach used in initial gathering of the data. You must consider the authenticity of the data and the source.

When collecting secondary data there are some important guidelines and these are outlined below. Most importantly, you need to plan your data collection and develop a strategy. Identify the type of data you need—is it numerical, textual or pictorial. For example, if it is data on the economies for different countries then the source of the data may be government treasury departments, if it was how women are portrayed in the media then library holdings of newspapers would allow an historical analysis of pictures of women or if it was how decisions relating to a project were made then obtaining minutes of planning meetings would be useful. Then decide how you are going to record the data—are you going to collate a bank of photographs or download Internet data into spread sheets or if taking information from paper-based sources then constructing a coding sheet would help.

Documentation is essential when collecting secondary data; you must record where the data is in terms of name and address of library, web address, etc., book and page numbers, etc. It is important too to record the date when you collected the data and the authority behind the data collection, for example, the UN or the Chinese government, etc. Make sure you read the notes about the data, how it was collected, who is included, what are the units/currency things are measured in, any special events and so on. Watch out for structural breaks. Perhaps the time series changes because the definition of the data changes. For example, perhaps in an unemployment data series after a certain date all those over a certain age are deemed to have retired rather than being unemployed. Perhaps part of a country has split from the original country to form as a new country. If this area was the principal producer of coal then the time series of coal production for the original country would show a marked decline after the area seceded from the original country.

So in collecting secondary data remember to PROD, that is, plan, read, observe and document! There are many sources of secondary data, notably government agencies, educational institutions, companies, non-profit making institutions, public and specialist libraries and the Internet. You will likely become a major user of the Internet in collecting secondary data, so some advice on Internet searching is presented.

If secondary data is used in student research then it is important to:

1. Justify why the secondary data is relevant to the research questions.
2. Understand how the data was collected.
3. Put effort into the analysis—after all the data collection was probably easier so to meet assessment requirements then one would anticipate more analysis. To give an idea of what might be done using secondary data sources, see for example the paper by Alverez et al. (2012), which illustrates how a student might use a secondary data source for research.

7.2 WEB SEARCH SKILLS

Until the availability of search tools became widespread, surfing was a typical approach for finding information on the web. Surfing is unstructured browsing, whereby links are followed from page to page and educated guesses are made along the way to arrive at the desired piece of information. Surfing is fun when there is time; however, it is an ineffective method. To promptly find relevant information from a variety of sources, knowledge of the available search tools is necessary. These generally fall into two categories: *browsing* through subject directories and keyword *searching* using search engines. This section aims to introduce you to efficient and effective use of these tools, how to select the best tool and approach for the task at hand and how to critically evaluate the performance.

Introduction to Search Tools

The term 'search engine' is often used generically to describe both true search engines and subject directories; however, the difference between them is how their listings are compiled. Subject directories are organised indexes of subject categories that allow you to browse through lists of sites by subject. They are compiled and maintained by humans, and pages are assigned categories either by the web page author or by subject directory administrators. Many subject directories also have their own keyword searchable indexes. Search engines, however, have their indexes created automatically by software known as 'spiders', 'robots' or 'crawlers' that travel the Internet to discover and collect resources. Searchers can connect to a search engine site and enter keywords to query the index, and web pages and other Internet resources that satisfy the query are identified and listed.

Subject Directories

These are compiled and maintained by humans, and so tend to be smaller than those of the search engines, which means that result lists tend to be smaller as well. Website owners/managers can ask to have their sites included in a subject directory, and less-selective subject directories will accept most submitted sites. Those with more stringent quality standards will be more selective and will more consistently lead to better-quality material. Subject directories are therefore best for searching for information about a general subject, rather than for a specific piece of information. The best known general subject directories are as follows:

Google	http://www.google.com
Yahoo	http://www.yahoo.com/
About	http://www.about.com/
Lycos	http://www.lycos.com/
Magellan	http://magellan.excite.com/
LookSmart	http://www.looksmart.com/
Open Directory	http://dmoz.org/

A subject directory contains an overview of subjects, subdivided into often quite broad categories such as art, recreation and science. Yahoo! is the one of the biggest and most popular programs, but covers less than 5 per cent of the web. Their category listings are shown in Table 7.1. There is no standard for such a system, and every subject directory uses its own categorisation.

For instance, the route to follow for information on the Scottish poet Robert Burns is as follows:

Arts & Humanities>Literature>Poetry>Poets@>Scottish Poets@

Select the link for Yahoo! Canada Directory>Poets> Robert Burns (1759–1796) and note the number of results returned.

In addition to browsing through the subject headings, most subject directories also provide keyword search facilities. This might create the impression that you are not searching a subject directory but that you are searching the entire Internet via a search engine. However, when you use the keyword search facilities of a subject directory, you are still searching those websites selected by the specialists in the subject directory.

Now go back to the Yahoo! homepage (either by the back button or simply retyping the web address: http://www.yahoo.com again) and type in the words 'Robert Burns' in Yahoo's search function field and press the search button.

You should notice from the screen, that you are provided with a larger number of results than when you clicked through the different levels of the previous headings (it is

Table 7.1 Category Listings

Arts & Humanities Literature, Photography…	**News & Media** Full Coverage, Newspaper, TV…
Business & Economy B2B, Finance, Shopping, Jobs…	**Recreation & Sports** Sports, Travel, Autos, Outdoors…
Computers & Internet Internet, WWW, Software, Games…	**Reference** Libraries, Dictionaries, Quotations…
Education College and University, K-12…	**Regional** Countries, Regions, US States…
Entertainment Cool Links, Movies, Humour, Music…	**Science** Animals, Astronomy, Engineering…
Government Elections, Military, Law, Taxes…	**Social Science** Archaeology, Economics, Languages…
Health Medicine, Diseases, Drugs, Fitness…	**Society & Culture** People, Environment, Religion…

SOURCE: Yahoo.com

also a much quicker way of searching!). Included in these results are matches from different categories, some of which you may not have considered browsing through.

Gateways and Subject Guides

Gateway pages or subject guides are web pages with many links covering a subject area, discipline or field, usually done by an expert in the field. A guide compiled by a subject specialist is more likely than a general subject directory to produce relevant information, and is usually more comprehensive than a general guide. Such guides exist for virtually every topic and some websites act as collections or clearing houses of specialised subject directories.

Search Engines

Search engines allow the user to enter keywords that are run against a database that has been created automatically by 'spiders' or 'robots'. All search engines operate on this principle but various factors influence the results from each. Database size, frequency of update, search capability and design, and speed may lead to amazingly different results. In most cases, search engines are best used to locate a specific piece of information, such as a known document, an image or a computer program, rather than a general subject.

Search engines have three major elements: the spider, the index and the search engine software. The spider (also called a 'robot' or 'crawler') visits a web page, reads it and then follows links to other pages within the site. The spider then enters its findings into the index (also called the catalogue), which contains a copy of every web page that the spider finds. The spider returns to the site on a regular basis and if a web page changes, then the index is updated with new information. The update does not always happen immediately, in which case the new information is not yet searchable. Finally, the search engine software is the program that sifts through the millions of pages in the index to find matches to a search keyword(s) and ranks them in accordance to its rules.

The main rules as to which are most relevant involve the location and frequency of keywords on a web page so that pages with keywords appearing in the title are assumed to be more relevant than others to the topic. Search engines will also check to see if the keywords appear near the top of a web page, such as in the headline or in the first few paragraphs of text, assuming that any page relevant to the topic will mention those words in any initial description. The other major rule is the frequency with which keywords appear, and those with a higher frequency are often deemed more relevant than other web pages. This is the basic principle but no two search engines will return the same results as some index more web pages than others, and some index web pages more often than others. Some search engines (and directories) will rank pages higher if it has links to the page, and some search engines rank pages with keywords in their meta-tags, but others do not. Popularity is another ranking criterion whereby pages that are linked from other pages are regarded as more relevant.

A selection of search engines is listed below.

AltaVista	http://www.altavista.com
Excite	http://www.excite.com
Google	http://www.google.com
HotBot	http://www.hotbot.com
FAST	http://www.alltheweb.com
Infoseek	http://infoseek.go.com
Northern Light	http://www.northernlight.com

As an example, go to the Google search engine to search for a poem by Robert Burns entitled 'To a Mouse'. In the appropriate search field, type in the words 'To a mouse by Robert Burns' and press return. Over 300 hyperlinked hits are obtained! Your screen should now display a number of results, detailing the page title (usually hyperlinked), the complete web address and a brief description for that page.

Advanced Searching

As the Internet is a vast computer database, searching its contents must be done according to the rules of computer database searching. Much database searching is based on the principles of 'Boolean logic'. This term refers to the logical relationship among search terms, and is named from the British mathematician George Boole. The commands of Boolean logic are given in the following sections.

The OR Command

The Boolean OR command is used to allow any of the terms you have specified to be present on the web pages listed in the results. It can also be described as a Match Any search.

1. Go to the main AltaVista page at http://www.altavista.com and search for the term 'ireland' (without the quotation marks). Write down the number of results matched.
2. Now go back to the main AltaVista page and search for the term 'eire', again writing down the number of results matched.

We shall now go on to use the OR command to search for both terms together. Go to the Advanced Search AltaVista page and search for ireland OR eire.
3. The result number should be significantly higher than the previous two results pages.

The AND Command

The Boolean AND command is used in order to require that all search terms be present on the web pages listed in results. It can also be described as a Match All search. You can also use the '+' sign directly in front of words that you want included in your results.

Now go back to the main AltaVista page and search for eire + ireland. The result number should be significantly lower than the previous three results pages.

The NOT Command

The Boolean command is used when a particular search term is *not* to be present on the web pages listed in the results. It can also be described as an Exclude search. You can also use '–' sign directly in front of words that you want excluded from your results.

Now go back to the main AltaVista page and search for eire–ireland. The result number should be lower still than the previous four results pages.

Additional commands that some search engines provide are given in the following sections.

The NEAR Command

This command is used in order to specify how close terms should appear to each other. That is, the terms may not be right next to each other, but very close. This command is often only available on advanced search pages.

Now go back to the main AltaVista page and go to the Advanced Search page. In the Boolean query box, type in peanut NEAR butter. This query would find return documents with peanut butter, but probably not any other kind of butter.

Note: The search engines listed below show the exact distance between terms searched using the NEAR command (this varies by each service).

AltaVista—within 10 words of each other
Lycos—within 25 words of each other
WebCrawler—within two words of each other

Phrases

Put quotation marks ' ' around a group of words and you've got a phrase. Only text with an exact match to the phrase will be returned. You can use the NOT command to exclude search terms.

Now go back to the Advanced Search AltaVista page, and search for natural disasters but excluding earthquakes by typing 'natural disasters' AND NOT earthquakes.

Nesting

Nesting allows you to build even more complex queries. You nest queries using parentheses (marking off specific sections of your search term by the use of brackets).

For example, try searching for the Boolean phrase impeachment AND (johnson NOT clinton) to find any web pages matching the term 'impeachment' but only dealing with Andrew Johnson.

Easier Advanced Searching

More and more search engines are now making advanced searching much easier and often you will not be required to use or remember any of the computer database search commands detailed earlier. For example, go to the Excite search page at http://www.excite.com and follow the link to the Advanced search, found at the foot of the page. You will notice that rather than a single text box for entering your query, there are four to begin with which you can increase as you wish. You enter a keyword or phrase into a box and then grade it according to importance, which replaces the need to use the Boolean commands. A particularly useful facility is the ability to narrow the search down to specific countries or domains. For instance, you may wish to search all sites within India, or perhaps only commercial (.com) or governmental (.gov), etc.

Meta-search Engines

Meta-search engines allow you to search several databases at the same time, via a single interface (e.g., they might trawl the whole of Webcrawler, Infoseek, Excite, Yahoo! and HotBot for you in one go). The great advantage is the breadth of search, but there are drawbacks also. A query submitted to a meta-search engine, with its uniform search interface and syntax, is to be applied against the diversity of individual search engines and so meta-search engines cannot take advantage of all the features of the individual search engines. Also meta-search engines generally do not conduct exhaustive searches and only make use of the top 10–100 ranked pages from each of them. A selection of meta-search engines is listed below.

Metacrawler	http://www.metacrawler.com/
Dogpile	http://www.dogpile.com/index.html
ProFusion	http://www.profusion.com
SavvySearch	http://www.savvysearch.com

Go to the Metacrawler page and you will be presented with a keyword search box.

Notice that there is some Boolean functionality provided, and that you can limit your search to UK sites and specific time periods. You should notice from the results page that meta-search engines provide the name of the search engine that provided the results.

Invisible Web

The following text is adapted from an article by Chris Sherman found at http://www.freepint.co.uk/issues/080600.htm#feature:

> There's a big problem with most search engines, and it's one many people aren't even aware of. The problem is that vast expanses of the web are completely invisible to general purpose search engines, and this 'Invisible Web' is in all likelihood growing significantly faster than the familiar 'visible' web. Search engines use automated programs called spiders or robots to 'crawl' the web and retrieve pages and rely on links to take them from page to page. Crawling is a resource-intensive operation, and puts a certain amount of demand on the host computers being crawled. For these reasons, search engines will often limit the number of pages they retrieve and index from any given website. These unretrieved pages are not part of the Invisible Web as they are visible and indexable, but the search engines have made a conscious decision not to index them.

Many of the major engines are making serious efforts to include them and make their indexes more comprehensive, but unfortunately, the engines have also discovered through their 'deep crawls' that there's a tremendous amount of duplication on the web. Current estimates put the web at about 1.2–1.5 billion indexable pages, but these have been distilled down to about 500 million pages. But these numbers don't include web pages that can't be indexed, or information that's available via the web but isn't accessible by the search engines. This is the stuff of the Invisible web.

The basic reason that some pages cannot be indexed is that there are no links pointing to a page that a search engine spider can follow, or a page may be made up of data types that search engines don't index. The biggest part of the Invisible web is made up of information stored in databases, and when an indexing spider comes across a database, it can record the database address, but nothing about the documents it contains. There are thousands—perhaps millions—of databases containing high-quality information that are accessible via the web, but in order to search them, you typically must visit the website that provides an interface to the database. The advantage to this direct approach is that you can use search tools that were specifically designed to retrieve the best results from the database. The disadvantage is that you need to locate the database in the first place, a task the search engines may or may not be able to help you with, but fortunately there are several reasonably thorough guides to the Invisible web, listed below. When using these searchable databases, keep your queries very broad.

WebData   http://www.webdata.com/

Lycos Invisible Web Catalog:

http://dir.lycos.com/Reference/Searchable_Databases/
Direct Search   http://gwis2.circ.gwu.edu/~gprice/direct.htm
The Invisible web   http://www.invisibleweb.com/

Specialist Search Tools

Most of the main search engines provide searching for a variety of specialised topics, for example, graphics, news, software or businesses and will always remain the most popular. There are also dedicated search engines catering to topics that either use their own (often huge) dedicated databases or search the web via other search engines.

News Search Engines

The services below are an excellent way to search for current news stories from hundreds of sources on the web. Since they crawl only news sites once or twice a day, the results are usually highly relevant.

Moreover   http://www.moreover.com/news/index.html
Excite Newstracker   http://www.bbc.co.uk

Mailing Lists

Subscribing to a mailing list is useful if you have a particular interest, but bear in mind that most of the information contained are opinions.

Topica   http://www.topica.com/
Liszt   http://www.liszt.com/
LISTSERV   http://www.lsoft.com/lists/list_q.html

Newsgroup Searching

This is a very useful resource for locating messages and newsgroups relevant to a specified topic. Bear in mind that these are peoples' opinions only, but can still be useful if you find someone with an interest in your own area that you can email with a question.

Deja.com   http://www.deja.com/usenet/

Answers' Searching

Answers' searching is when someone is looking for the answer to a question, rather than a specific website. This resource has catalogued previously asked questions.

Ask Jeeves   http://www.ask.co.uk/

Financial/Business Searching

FinancialFind   http://www.financialfind.com/
FML Exchange   http://www.fmlx.com/

Searching Tips

- Type the most important term in your search first.
- No search tool is fully comprehensive, so choose an appropriate one from three main types: directories, search engines and meta-search engines.
- Good websites have a 'related links' section, which list other websites dealing with similar or related topics. It is always worth browsing these sites as they have been evaluated to a degree, and can therefore save you time wading through your own search results.
- The more specific your criteria, the better the results. Avoid general terms and words with common synonyms, and use the advanced search feature.

- If you don't know enough fine detail about your chosen topic, be curious—find a general site, and once you have discovered more about the topic, embark on a more accurate search.
- If you know exactly what you are looking for, it might be worth guessing the web address first of all.
- Searching can be time consuming, don't even try to read everything you come across. When you access a page, scan it quickly and be ruthless in abandoning it if it doesn't look useful.
- If you're unsure of the spelling of the term you wish to search for use wildcards. Do this by using an *, for example, search for 'organi*'.
- Major search engines are constantly changing and updating their facilities. Try to keep up with their advancements as the new features are designed to help improve your searching.
- Remember to use the bookmarking function—don't rely on your memory!

Online Encyclopaedias

Many academics express concern about the use of these—partly because many of them such as Wikipedia rely on users to provide and update content—which might not be correct and worse it might deliberately be biased and incorrect. For this reason, students are sometimes penalized for quoting Wikipedia. Personally I think that the likes of Wikipedia are very useful—it gives an overview and provides references. One should be critical and cautious of these sources but I am sure they are part of the future. A detailed discussion on Internet research can be found elsewhere (Hewson et al. 2003; O'Dochartaigh 2001).

7.3 EXERCISES

Exercise 1

Compile a list of the format of secondary data that physically exist at your nearest library. Comment on the usefulness of each format in terms of your own area of research.

Exercise 2

Identify a recent journal article written by an academic or researcher. Note all the references used in this particular article and use a catalogue or search engine to locate them.

- What proportion of them can you obtain from your own library?
- What proportion can you obtain by downloading from an Internet site?
- What proportion can you not find any reference to, or would be prohibitive to obtain?

Exercise 3

Read the notes 'WWW search skills' and carry out the exercises included in the text.

7.4 REFERENCES

J. Alverez, J. Canduela and R. Raeside, 'Knowledge Creation and the Use of Secondary Data,' *Journal of Clinical Nursing* 21 (2012): 2699–2710.

C. Hewson, P. Yule, D. Laurent and C. Vogel, *Internet Research Methods: A Practical Guide for the Social Sciences* (London: SAGE Publications, 2003).

Human Development Report, *Beyond Scarcity: Power, Poverty and the Global Water Crisis* (New York: UNDP, 2006), http://hdr.undp.org/hdr2006/statistics/

N. O'Dochartaigh, *The Internet Handbook: A Practical Guide for Students and Researchers in the Social Sciences* (London: SAGE Publications, 2001).

D.W. Stewart and M.A. Kamins, *Secondary Research: Information Sources and Methods* (London: SAGE Publications, 1992).

CHAPTER 8

Surveys

8.1 INTRODUCTION

In business and management, data is frequently collected through surveys. The survey method involves asking individuals questions face to face, by telephone or via questionnaires of individuals, and departments or companies to find out personal, company or sector information. Surveys are conducted on a continuum ranging from a small scale to a large scale of population; sometimes the whole population is involved, which is called a census. The principle is to collate answers to a number of questions and this lends itself to a more quantitative approach in terms of data analysis. The critical issue with surveys is the representativeness of those who you survey. In many surveys, researchers feel satisfied if 20 per cent of people respond to their survey.

So, what can we say about the remaining 80 per cent? It all depends on how representative our respondents are! Many researchers, largely because of convenience, end up conducting questionnaire-based surveys. However, bear in mind that the approach is overused and people and organisations are tired of surveys and as a result response rates are poor. Moreover, this questions the reliability and validity of the findings. The process of questionnaire surveying is depicted in Figure 8.1.

Surveys are well explained in many research methods books and you should consult relevant chapters in Bourque and Fielder (1995), Bryman and Bell (2011), Blumberg et al. (2011) and Ghauri and Gronhaug (2002).

To some extent, regardless of the type of research followed, many students of business and management and social science end up doing some form of survey. Many are successful, but in our experience many fail to collect data well, and this compromises future work. If the survey is not well designed and formulated then you may well face the criticism of 'garbage in, garbage out' when you analyse and report on your research.

Figure 8.1 The Survey Process

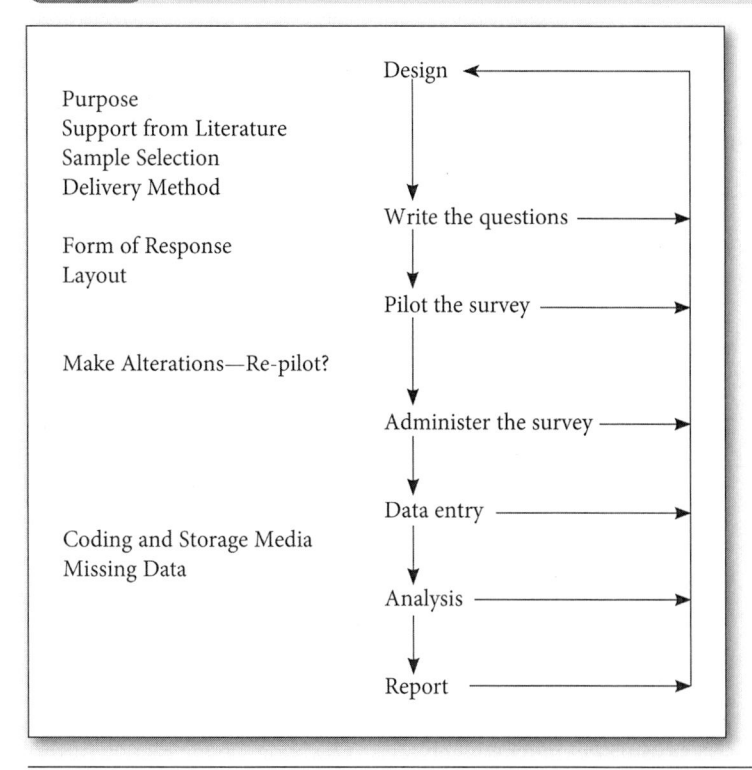

```
                            Design ◄─────────────────┐
Purpose                       │                       │
Support from Literature       │                       │
Sample Selection              │                       │
Delivery Method               ▼                       │
                            Write the questions ──────┤
Form of Response              │                       │
Layout                        │                       │
                              ▼                       │
                            Pilot the survey ─────────┤
                              │                       │
Make Alterations—Re-pilot?    │                       │
                              ▼                       │
                            Administer the survey ────┤
                              │                       │
                              ▼                       │
                            Data entry ───────────────┤
Coding and Storage Media      │                       │
Missing Data                  │                       │
                              ▼                       │
                            Analysis ─────────────────┤
                              │                       │
                              ▼                       │
                            Report ───────────────────┘
```

SOURCE: Authors' own.

Thus, it is important to go through the steps in Figure 8.1; this is the subject of the ensuing sections.

8.2 DESIGN

First, but not to put you off, is a survey really needed? All surveys, except perhaps for government-enforced surveys such as demographic censuses, suffer from low response rates, which mean that there is always the question, 'can we really say anything about those who do not answer?' So from the point of view of a student dissertation or any other study we tend to advocate that surveys are the last resort and are meant for those who lack imagination. Nevertheless, the majority of business and management students seem to end up doing surveys. But before doing this, please think whether secondary data—either statistical compilations or existing surveys—could answer your research questions.

The choice of design depends on what is to be answered and the depth one needs to probe. If it is straightforward information that is required then surveys can be very simple and can comprise questions about, say for instance, the investigation of a tourist visitor attraction. The questions could be:

- How often have you visited?
- Approximately how long did you spend there on your last trip?
- Approximately how much money did you spend?

But if you wish to find out about the person's experiences, feelings and attitudes then the survey design becomes more complex.

The general design principles of a survey are:

- Keep it short.
- Open layout.
- Clear, short, unambiguous questions.
- Imbedded instructions—give examples.
- Have scales all going one way.
- Have a simple return mechanism.

To design a survey well, the purpose has to be carefully formulated. Try to think of questions as variables, that is, responses that vary between individuals (or cases in statistical terms). Ensure you include a dependent variable or response variable and an independent variable. For example, if you were interested in the contribution of quality management to revenue, make sure you ask what the company's revenues were and think how this will be answered. Will it be total company revenue or division revenue? Questions will have to be asked to ascertain the effort put into quality management. These could be:

- Estimates of staff time
- Direct costs of materials
- Number, frequency and duration of meetings
- Some derived measure from the amount of non-compliances and the degree to which the company is customer centric

In addition, one needs to know about other factors may affect revenue, such as the degree of competition, retail price index, purchasing power, company size, number of competitors and so on. These are variables which are not directly related to the research question or hypothesis and are called *control variables*; often these far outnumber the other variables. The situation is illustrated in Figure 8.2.

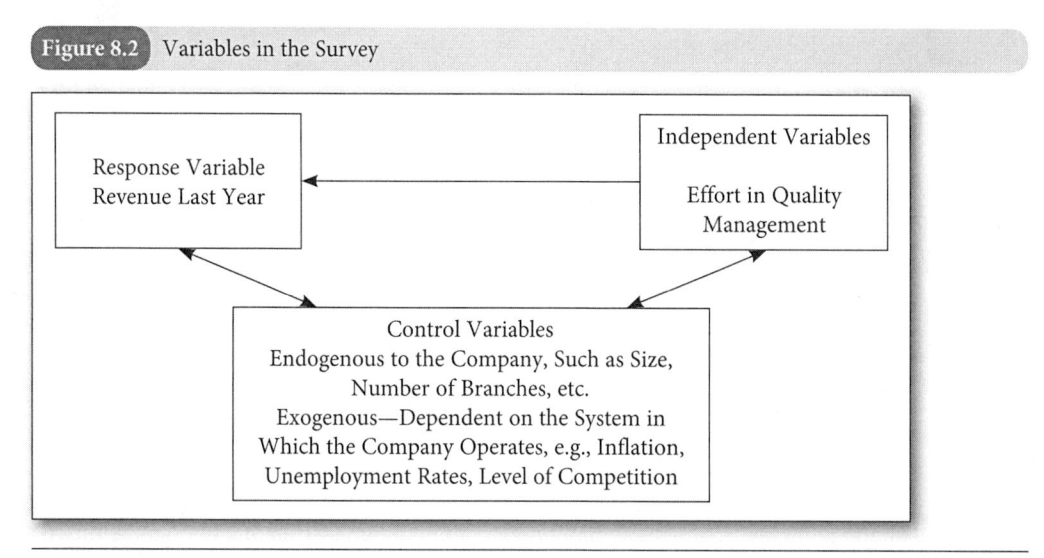

Figure 8.2   Variables in the Survey

```
┌─────────────────────┐         ┌─────────────────────┐
│ Response Variable   │ ←────── │ Independent Variables│
│ Revenue Last Year   │         │ Effort in Quality    │
│                     │         │ Management           │
└─────────────────────┘         └─────────────────────┘
         ↘                             ↙
         ┌────────────────────────────────────────┐
         │         Control Variables              │
         │ Endogenous to the Company, Such as Size,│
         │         Number of Branches, etc.       │
         │  Exogenous—Dependent on the System in  │
         │ Which the Company Operates, e.g., Inflation,│
         │ Unemployment Rates, Level of Competition│
         └────────────────────────────────────────┘
```

SOURCE: Authors' own.

The questions included in the survey should be relevant to the research aims and must be culled from the literature. Study of the literature helps with the choice, form and wording of questions. In some areas, such as Human Resource Management or Marketing, this is very important. For example, how does one measure motivation or satisfaction? Typically, a number of questions (often referred to as a bank of questions) are asked to measure these intangible concepts. It is most unlikely that your survey will be successful unless you consult the relevant literature and, typically, most surveys are adoptions of ones which have been run in the past.

From the literature you can find out how others have surveyed the area, the types of questions asked, how problems have been overcome and the type of survey that has been used.

The survey must be directed to the relevant population and this takes careful thought. Who has information you require—customers, suppliers, owners, current or ex-employees? For example, if one was doing research on how images of a country's heritage influence tourist choice, one could speculate to approach a middle-aged affluent subgroup, as it could be assumed that these people are more likely to engage in tourism to experience the heritage of that country. Think carefully about the target group of your survey. Are they in a position to be able to answer the questions? Once the relevant population has been decided then the principles of sampling outlined in Chapter 5 should be followed.

Regarding the choice of the survey delivery platform, one can choose between face-to-face interviews, mail questionnaires, peer dissemination, telephone and email/web questionnaires. The choice is for you, the researcher, to make. The delivery mechanisms are compared in Table 8.1.

Table 8.1 Survey Delivery Platform

Delivery Platform	Who Completes—Surveyor or Surveyed	Cost	Length of Questionnaire	Response Rates	Comments
Household	Either	Expensive	Can be long	High	Not advisable for a student approach—safety issues and too time consuming, no compulsion for householders to reply.
Street	Surveyor	Medium	Short	Medium	Concern as to the validity of responses and those surveyed tend to be self-selected—so are they really representative?
Telephone	Surveyor	Medium	Short	Depends on culture	In the UK, student-based telephone surveys of individuals do not appear to illicit a very high response rate but in Hong Kong they do. Can work for company-based surveys—but little detail is given and not of a confidential nature.
E-mail/Web	Surveyed	Cheap	Short	Low	Reliability concerns—can be multiple responses. Companies pay little attention to these.
Mail	Surveyed	Cheap	Variable	Low	For individuals need a reply paid envelope—not required for companies.
Customer	Either	Medium	Medium	Variable	Usually requires permission of retailer.
Part of an institution or organisation	Surveyor	Cheap	Medium	High	Can have high response rates if the surveyed think that it is advantageous to them. But can be biased responses as the respondents try to portray their position to their superiors.

SOURCE: Davies et al. (1993).

8.3 QUESTIONS

The questions asked in a survey must be clear and unambiguous. Remember that the person being surveyed will not understand your intentions if the questions are unclear. Moreover, the meaning of the questions must be absolutely clear, especially if you are not present while the respondents are taking the questionnaire. Be careful to include questions which only ask one thing. For example, how would people answer the following question: Do you approve of fast food advertising and what is the most suitable media?

Questions should not lead the respondents to the response. The question 'Most consider that membership of the World Trade Organisation helps national development, do you agree?' is clearly leading the respondent and so is a biased form of question.

Open and Closed Questions

Open questions are questions in which respondents are asked to describe issues or state their views and feelings. Answers are given as textual statements. Analysing these answers is both time consuming and difficult, especially, if the respondents deviate from the question or misinterpret it. In addition, some respondents are put-off by open questions as they find writing prose difficult or get concerned about grammar or spelling.

Closed questions restrict the choice available to the respondents but often the respondents find these easy to deal with. A comparative summary of open and closed questions is presented in Table 8.2.

Examples of closed questions are now given below:

1. *Checklist*: where the respondent is asked to tick appropriate responses such as reasons for an action or things experienced. For example:

What was most important to you in applying to the university in which you study?

1. Reputation
2. Accessibility
3. Costs
4. Subject
5. Delivery

In this case, this question allows only one item to be selected but other questions can be asked that allow multiple responses. This type of question is popular and fairly easy to analyse. However, it is difficult to think of all the categories.

2. *Ranks*: Respondents are asked to rank items in a given list in order of preference or importance. For example:

Table 8.2 Open and Closed Questions

Question Type	Advantages	Disadvantages
Open	Allows freedom of expression in conveying views	Time consuming
	Allows categories/cases to be established by the respondent	Answers have to be coded
	Allows meaning to be clarified by the respondent	Greater effort required by respondents
	Allows for more depth	Misinterpretation
	Allows for more exploration of the area	Hard to read
Closed	Easy to process answers	Loss of spontaneity of respondent
	Enhances comparability between cases	Difficult to make forced choice answers or mutually exclusive responses
	Clarify meaning to respondent	Difficult to make forced choice answers or exhaustive responses
	Completion is easier	Wide variation by respondents in interpretation of forced choice answers
	Reduces variability in analyst's interpretation	Can be irritating
		Makes questionnaire long

SOURCE: Davies et al. (1993).

Rank in order of importance for choice of university being most important.

1. Reputation
2. Accessibility
3. Costs
4. Subject
5. Delivery

Again, analysis is straightforward, but it is difficult for people to rank more than five items and poor selection of options on the list can compromise the research. Sometimes the list is formed from open questions given in a pilot survey in which respondents are asked to list the three most important things. A problem with ranks is that if 'a' is preferred to 'b' we don't know by how much it is preferred. Also, ties can be difficult to handle.

3. *Distribution*: Often in marketing studies the respondent is asked to distribute points or factitious currency between the options to reflect importance. For example:

Distribute 100 points across items, as to why you chose this university, where the highest value is given to the most important.

1. Reputation
2. Accessibility
3. Costs
4. Subject
5. Delivery

This approach allows scale to be represented in the analysis but respondents often find trouble allocating the points and feel that a high degree of precision is required.

4. *General data questions*: Such as approximately how much do you spend on food per week? Or what is your age? These are easy to answer and analyse.

5. *Categories*: Many researchers feel that respondents do not like giving age or approximate salary or revenue information so categories are given and the respondent is asked to tick the appropriate box. For example:

What age group (in years) are you in?

11 to 16   17 to 25
26 to 60   60+

We do not think that category questions are as good and straightforward as the data collection questions, as respondents can get annoyed if, say, they have just had their 26th birthday and are put into a higher category. In addition, the categorisation has to be carefully chosen, especially if one is comparing data to secondary data—say a census or other publicly available data. Often no matter how much care is taken, category end points do not meet up between surveys.

6. *Likert scales*: Here respondents are asked to rate a response on a four-, five- or seven-point scale. For example:

How important is the reputation of the university to you? Please tick.

Very unimportant [ ]   Unimportant [ ]   Neutral [ ]   Important [ ]   Very important [ ]

These questions are easy to answer and analyse. Some researchers argue for even numbered scales so that the respondent is forced to choose between negative and positive. But we think that in most business and management issues respondents are asked questions that they genuinely feel neutral about. We recommend odd-numbered scales.

7. *Attitude statements*: This is similar to the above and respondents are given a list of statements and asked for their level of agreement. For example:

Strongly
Disagree [ ]   Disagree [ ]   Neutral [ ]   Agree [ ]   Strongly
Agree [ ]

Fees are too high?

8. *Semantic scales*: Here the respondent is asked to mark his/her strength of feeling or opinion on a line which goes from low to high. The researcher then gives the position on the line a mark, usually out of 10. This can be done with a ruler, but it is our experience that judgement is sufficient. For example:

Mark your view of the difficulty of this subject on the following line:

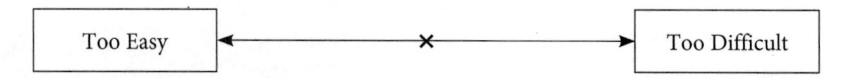

| Too Easy | ←———×———→ | Too Difficult |

Here the individual would score 6.

You must ensure that all the scales go from negative to positive or small to large—this seems to be how people think. Some researchers suggest reversing the occasional scales to ensure that the respondent is paying attention. We do not recommend this and think this leads to confusion and annoys the respondent.

The 'Don't Know the Answer'

In many surveys, questions have a 'Do not know' box or 'Not applicable'. Care should be taken here as the inclusion of 'Do not know' or 'No comment' may be giving respondents an escape route to avoid answering the question. This is a particular concern for sensitive questions.

Harmonisation

This is the process by which questions are kept the same between surveys. Therefore, this process allows comparisons of questions across surveys. The focus is on developing standardisation of categories and definitions. Any secondary sources that you may wish to use now ensures that the categories such as age and spending and any definitions (such as definitions of small medium-sized enterprises) are compatible with the secondary source—this must be done at the design stage. For UK definitions and categories, see harmonised questions at www.statistics.gov.uk/harmony.

Utilising harmonisation is a good practice and should help minimise the time and costs in survey development and help validation by using tested forms of questions. By allowing comparison to secondary sources, you will be able to get a lot more out of your survey and research.

Questionnaire Layout

Here are a few pointers to understand the questionnaire layout:

1. Have lots of 'white space' (empty space), this seems to ease the respondent.
2. In general ask easy non-contentious questions first and leave any questions which may damage the response rate until near the end.
3. Use 'go to' to avoid subjects completing irrelevant areas.

Ethics

In all research, ensuring integrity by following good ethical procedures is very important; conducting surveys are no exception. Key ethical issues are not to endanger participants—to help prevent this informed consent is necessary. This means that the purpose of the questionnaire must be clearly described to the potential respondent and the respondent has the right not to complete the survey; moreover, they should be able to withdraw at any time. Another feature is that in many cases confidentiality of respondents is required, this you need to ensure. In addition, after your work has been written up and assessed, data collected by questionnaire should be destroyed.

8.4 PILOT SURVEY

It is important that all surveys be tested before the actual survey is conducted. This is done to ensure that the questionnaire is clear to respondents and can be completed in the way you wish. Thus, it should be piloted on the sample to be used. It is best to observe respondents filling in the questionnaires to understand what questions trouble them. Then they are interviewed on what they thought about the questionnaire, how it could be improved and if anything is missing. If all is reasonably well then the sample can often be incorporated into the final database.

The pilot survey tests the following:

- Procedures for gaining informed consent
- Wording of the questions
- Sequence and layout of the questionnaire
- Fieldwork arrangements
- Analysis procedures

Additionally, the pilot survey can be used to train any fieldworkers you use and to help estimate both response rates and completion times.

Remember to document the pilot process, what you have learned and what adjustments have to be made. You should refer to the pilot process in your dissertation.

Presented here is a checklist for a pilot survey:

- Is it needed?
- Are there to be multiple ones?
- Procedure—how different is the pilot from the final?
- Are the samples representative—should they be?
- Consider
    - Were questions understood?
    - Were instructions clear?
    - Do the administrators understand?
    - Is the order correct?
    - Is the data likely to be meaningful and is sensible analysis possible?
    - What were the costs?
    - How long did completion take?
    - How were the administrative arrangements?

8.5 ADMINISTERING THE SURVEY

The administration of the survey can have a great impact on the response rate and the quality and reliability of the respondents' answers. In this section comments are made in regard to the administration of mail and e-surveys. The various ways of administering the different types of surveys are detailed next.

Mail Surveys

In order to ensure responses from the respondents, prepaid return envelopes should be included along with the survey questionnaire. Often, incentives are given, such as inclusion in a prize draw. However, for a student survey we have found that this is not needed, and can even be detrimental. If a company is going to complete the questionnaire then respondents will return the questionnaire regardless of whether there is a prepaid envelope or not. The following are some considerations for a mail survey:

1. It seems best to contact the respondents in advance and ask them to complete the questionnaire when it arrives.

2. Send the questionnaire out in 'waves', for example, send out, say 100, and two weeks later another 100, and so forth.
3. Reminders should be sent out 10 days to two weeks later. For a student project we recommend only one reminder be sent.
4. Code questionnaires to correspond to where they are sent so that you know who has replied and who is to be reminded.

E-surveys

Surveying by email or over the web is becoming more feasible; however, response rates tend to be very poor and one is uncertain as to who is actually replying—emails sent to senior managers may well be answered by their secretaries! Dillman (1999) argues that this can be helped by utilising a respondent-friendly design. This should take account of the inability of some users to receive and respond to web questionnaires with advanced features, account for both the logic of how computers operate and people's expectations of questionnaires. This really means that it is best to keep the electronic version as simple as possible. In addition, you need to remember that the respondent is doing other things, so avoid the need to save and restart. One or two small question surveys are, however, feasible by SMS text.

The technology is now readily accessible for web-based surveys, and there are some very good commercial survey providers such as Survey Monkey (www.surveymonkey.com/) and QuestionPro.com (www.questionpro.com). Should you embark on an e-survey, ensure you do read Dillman's *Email and Internet Surveys: The Tailored Design Method* (1999) and adhere to his list of the following principles of e-surveying:

1. Introduce with a welcome screen that is motivational, emphasises ease of responding and provides clear instructions.
2. Give computer operation instructions at the beginning of the questionnaire.
3. Provide specific instructions on how to respond to questions and submit the questionnaire—assume little or no computing knowledge on part of respondents.
4. Begin with a fully visible question.
5. Present questionnaires in a conventional form—that is normally used in paper questionnaires.
6. Limit line length to width of a standard screen.
7. Do not require user to answer a question before moving on to the next.
8. Construct questionnaires so that users can scroll from question to question—so they can quickly skip but still see questions that they do not need to answer.
9. Try and ensure that all option choices are visible on one screen.

10. Use graphical symbols or words that give a sense of where the respondent is in the completion process.
11. Be cautious about using question structures that have known measurement problems.

A major advantage of web surveys is that the responses are collated and can be easily downloaded into spreadsheets such as Microsoft Excel or even statistical packages such as SPSS. This helps reduce transcription errors. Remember that ethics procedures apply to e-surveys as they do to paper surveys.

We strongly recommend that questionnaires sent by e-mail be in the body of the text and not sent as attachments, as respondents have to save and then reattach the questionnaire. Often blank questionnaires are sent back. The best option is to use a web-based survey in which the web address is sent to potential respondents as an email.

In many web-based surveys the company used will, for a fee, analyse your data—this we do not recommend as it is often very unsophisticated—unless you pay a lot, you deny yourself a learning opportunity to learn about your respondents and how to analyse them and importantly you lose the opportunity to gain marks for the analysis you undertake.

Social Media

Many students make use of social media such as facebook to disseminate their questionnaire and receive completed questionnaires. This does work in terms of getting a reasonable sample size quickly. However, this approach is flawed as it tends to be unrepresentative, often going to the researchers' friends or to similar people. This means that the respondents come from a narrow cross-section and might well not be representative. This narrowing of the sample to similar people is also a problem with snowball sampling, in which one respondent refers the researcher on to another. As a researcher you need to debate if the convenience of using social media outweighs the sampling problems. A conversation on this is needed with your supervisor.

8.6 ENSURING HIGH RESPONSE RATES

Success with surveys often depends on good response rates, which is the number of respondents divided by the number approached. (The valid response rate is the number of responses which are suitable for analysis divided by the number approached—this should be the measure that is reported.) To ensure a high valid response rate, try to follow these guidelines:

1. Make the survey interesting to the respondents and convey this interest to them in a brief covering letter or statement.
2. If feasible, contact the respondents before the survey to ask for cooperation and impress on them the importance of the survey.
3. Keep the questions simple and the layout clear.
4. Only ask for information you require and which is of direct relevance to your research.
5. Ask for approximates or estimates rather than exact answers—if the respondent has to leave the question to find the answer from records, etc., then they are unlikely to return to the survey.
6. Do not ask the respondent to do calculations—if calculations are required, ask for the inputs and you do the calculations.
7. Do not ask for information which is readily available in the public domain.
8. If you seek confidential information then make sure you keep it confidential.
9. Send reminders out no later than two weeks after the initial sending of the questionnaire.
10. Make it clear that you are a student—many people feel sympathetic towards students and want to help—so often the response rate towards students is higher than those, say, of market research companies!
11. Ensure the questions have been piloted and are understandable.
12. Ensure that the survey is free from typographical and grammatical errors.
13. Make the survey look professional—use good-quality paper and a good printer.
14. Put the return address on the questionnaire usually at the end—as often a covering letter will become detached. It is probably best to organise the collection of the completed questionnaires.
15. Use the following checklist:

    - Good covering letter.
    - Follow up individuals—stress their value—sometimes respondents wait for a second approach, feeling that that is proof the questionnaire is important.
    - Keep questionnaires short and clear—final reminder could be a cut-down version of critical questions only.
    - Clear instructions.
    - Questionnaire does not appear bulky.
    - Restrict open questions.
    - Explain purpose and value.
    - Ensure interest to the respondent.
    - Length kept to a minimum.
    - Design/layout is appealing.
    - Minimise complexity—is help needed to complete—will respondent have information on hand?

- Leave plenty of 'white space'; do not crowd the pages—best to use more pages than have a densely packed page.

8.7 MISSING INFORMATION

Missing information from the surveys can limit your analysis. The most serious is 'unit non-response' and this is when a particular group does not reply. The unit could be one community, one sector of companies or a grade of employees. Thus, one can say nothing about these units as they are not represented. This will mean that you will need to either reconsider your initial plans as some comparisons cannot be made or resample for targeting the missing units.

'Item non-response' is when a question or some questions are not answered. This tends to be less serious than 'unit non-response' as often you have some information about the subject and sometimes their response can be inferred from how others have responded. This is called *Imputation*. This can be 'Informed Imputation' where you have a good idea of what the response would be. For instance, the questions were completed by interview and the respondents refused to give their age. However, through observation you could give an informed guess about the age. Another form is 'Hot Deck Imputation' in which the idea is to look through the data matrix for respondents who are similar to the one who has a missing item(s) and use the average of the similar respondents' responses to the missing item(s). Finally, there is 'Model-based Imputation' where statistical regression or neural network procedures are used to predict the missing values. If you impute values you must clearly state in any documentation that this has been done and how you have done it.

Generally, at undergraduate or postgraduate dissertation research you would not be expected to use imputation, but if you decide to then please consult Little and Rubin (2002), and Carpenter et al. (2006).

What Can Go Wrong?

For good criticisms of surveys, see Futrell (1994). Futrell gives 10 reasons why surveys fail; he groups these into problems with selecting the sample and the content and analysis of the survey.

Futrell's 10 reasons for why surveys fail are given below.

1. *Failing to use statistical sampling methods*: This leads to a biased design and directly weakens the validity of your research.

2. *Ignoring non-response*: If this happens then the representativeness of your work will be questionable as the non-respondents may have a completely different attitude to those who responded.
3. *Failing to assess the reliability of the survey*: Was the survey largely down to your personal influence, and if it was then it would be unlikely anyone could replicate your findings and thus your work cannot be taken seriously.
4. *Treating perceptions as objective measures*: Often one's intangible feelings about objects, concepts or people cannot be objectively measured and one should bear this in mind when drawing inferences from Likert scales.
5. *Treating surveys as an event and not a process*: Surveys are snapshots in a time continuum and at best represent that time period. So take care in inferring from history or making predictions when analysing surveys.
6. *Asking non-specific questions that can be interpreted in several ways*: This limits the inferences you can make and can render the conclusions invalid.
7. *Failing to ask all the questions*: There are many pertinent questions that you will only discover after the analysis has been completed.
8. Using incorrect or incomplete data analysis methods.
9. *Ignoring some of the results*: This is an unethical practice. Data may be excluded if you do not think they are representative, but you must clearly indicate this and the reasons for excluding the data.
10. Using the results incorrectly.

8.8 CODING AND DATA INPUT

It is recommended that questionnaire responses be entered into a spreadsheet such as Microsoft Excel. The first row should comprise headings, which identify each question or sub-question, each following row is a case, that is someone's responses. Each column is the answer to each question or sub-question. This is the conventional format for statistical analysis and the file produced is referred to as the Data Matrix. There is no need to do anything to general data collection questions. However, for questions which give a list of categories, the different categories should be given numerical codes. For example:

For the question, 'Do you agree?' Yes or No—code 'Yes' as 1 and 'No' as 0.
For gender code, 'Females' 0 and 'Males' 1. Note here no order is implied by the numbers.
For the question, 'What best describes your employment level?' (Manual unskilled, Manual semi-skilled, Manual skilled, Clerical, Junior management, Senior management, Professional); employment level would be coded 1–7.

The coding should be done on a master copy of the questionnaire and carefully documented.

Once the coding plan is complete, the results should be entered on a spreadsheet such as Microsoft Excel. Each row is a case, that is, an individual response produced on that individual's questionnaire. Each column is a response to a question, or a unit of a question if multiple responses are permitted. Excel data from spreadsheets can easily become input in analysis packages such as SPSS or SAS. Most students find data inputing and editing easiest in Excel.

If questions are unanswered or are missing then leave the cell blank—do not use '0s' or '99', etc. The first row is normally reserved for variable names and the first column for respondent ID.

For example, consider the short questionnaire given to six companies:

1. Which sector do you operate in?
   Manufacturing   *1*
   Commercial services   *2*
   Public service   *3*

2. How many employees work in your building? *Enter Answer*
3. Does your organisation have quality certification, yes or no? $Y = 1$
   $N = 0$
4. What are your views on quality management?

The coding for Question 4 is shown in Table 8.3.

If the first six respondents were entered into Excel, then the spreadsheet shown in Figure 8.3 would be formed; this is sometimes referred to as the *data matrix*.

Table 8.3 Coded Responses for Question Views on Quality Management

	Strongly Disagree	Disagree	Neutral	Agree	Strongly Agree
Quality management is time consuming	1	2	3	4	5
Quality management improves productivity	1	2	3	4	5
Quality management is expensive	1	2	3	4	5
Quality management is motivational	1	2	3	4	5

SOURCE: Authors' own.

Figure 8.3 Example of Data Entry into Excel

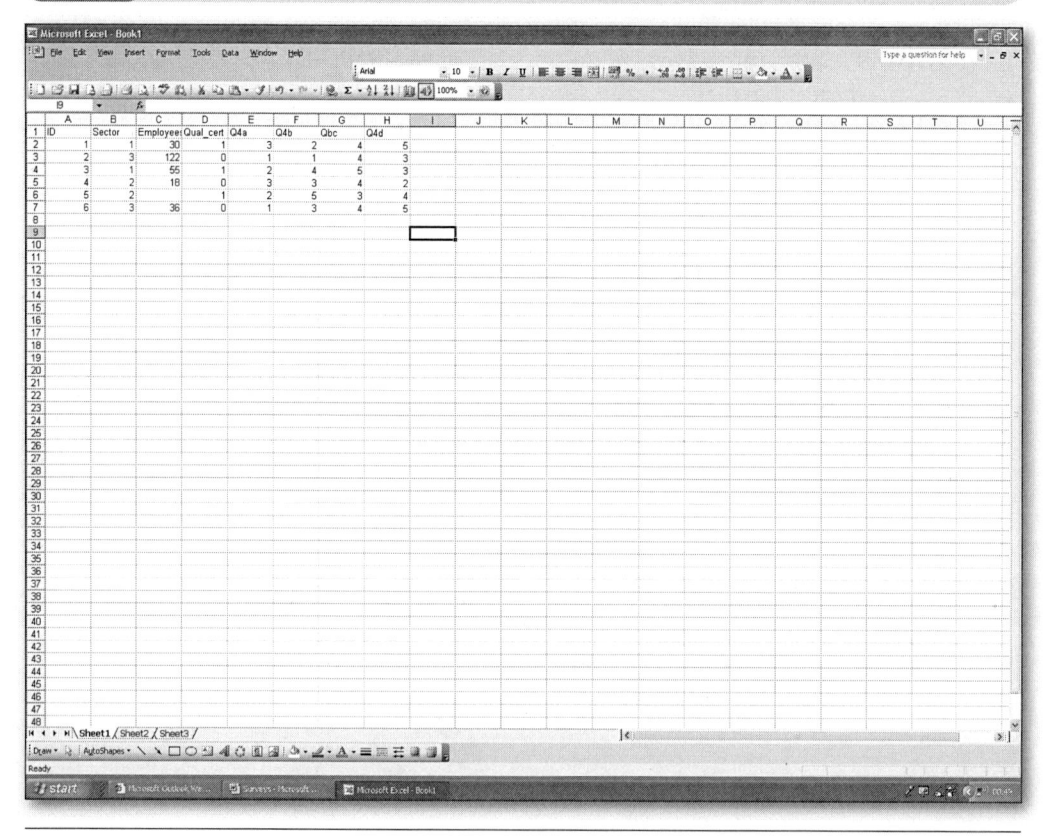

SOURCE: Authors' own.

8.9 GUIDELINES

To conclude this chapter we present some general guidelines for conducting surveys.

1. Questions asked need to be very simple and concise.
2. Do not overestimate the knowledge and ability of the respondent. (Try to avoid embarrassing them.)
3. Ensure everyone interprets the questions the same way.
4. Each question can only deal with one dimension or aspect. For example, the question, 'Do you enjoy studying Biology at University X?' The answer to this does not contain much information. Does a 'No' answer mean that they do not like the subject or University X?

- Questions need to be specific, not like, 'Is the workload on this module similar to other modules?'

5. Avoid leading or suggestive questions. For example, 'Degree level education is a good idea. Isn't it?'
6. Formulate questions in soft and polite language. Language used should be simple and straightforward. For example, 'Do you have notarised educational qualifications?' Such a question can create confusions in the respondents' minds.
7. If asking about concepts like 'satisfaction', also ask how important the issue under consideration is.

- Place questions in the 'right' order, that is from easy to complex and general to specific.

8. Avoid long passages of text—give the questionnaire an open feel.
9. You must pre-test the questionnaire and give time to conduct critical evaluation.
10. Avoid the trap of asking questions to get data, which might be useful but is of little relevance to your research—this makes the questionnaire unnecessarily long.
11. Make the return route clear for completed questionnaires.

8.10 SOCIAL NETWORKS

This is an emerging approach to social science and business research and is used to trace out patterns of communication, understand how individuals influence one another, how organisations partner with one another and how information flows in an organisation. See Cross and Parker (2004), Scott (2005) and Valente (2010) for more information. In this, the focus of analysis is the link between actors rather than the attributes of the actor (both link and attributes can be combined, see Gayen and Raeside [2010]).

To collect data requires that you find out about a person's or organisation's contacts and sometimes the frequency and importance of the link. A typical social network question is displayed in Table 8.4.

Giving this to a class of 10 students might yield the data shown in Table 8.5.

One can then compute the strength of the contact by multiplying frequency by value. One then can produce a matrix to represent the contacts as shown in Table 8.6. This matrix can then be copied and pasted into a social networks package such as UCINET 6 (Borgatti et al. 2002). This software can be obtained from Analytics Technologies (www.analytictech.com/). The file is pasted into the UCINET spreadsheet

Table 8.4 Social Network Question—List up to Three People Who You Have Discussed Mobile Phone Contracts with

Name	Frequency of discussion 4 = daily, 3 = weekly, 2 = monthly, 1 = less frequently	Value of discussion 1 = no value to 5 = high value

SOURCE: Authors' own.

Figure 8.4 Sociogram of Students Discussing Mobile Telephone Contracts

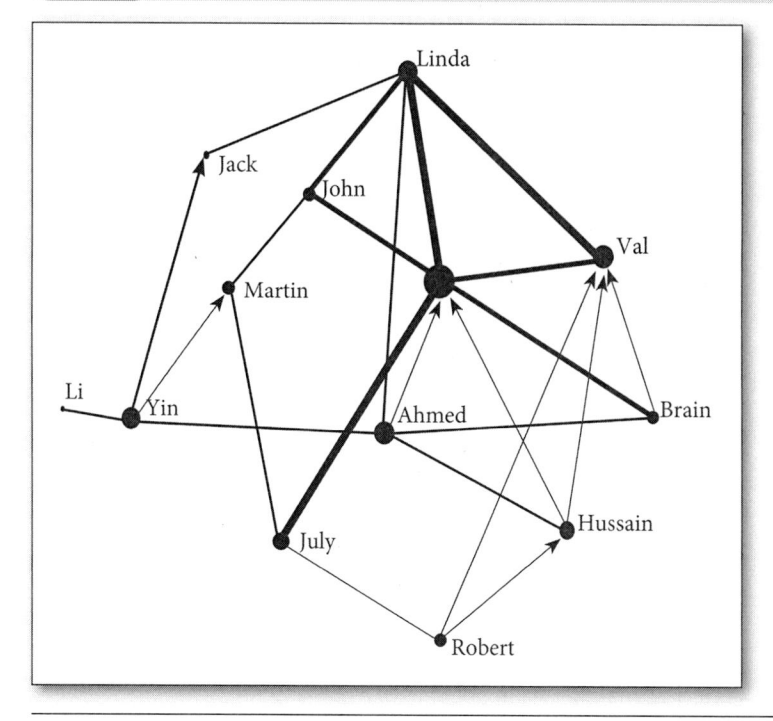

SOURCE: Authors' own.

and saved with the extension saved with the extension.##h. One can then visualise the pattern of connections, which is called the Sociogram; for this example, it is displayed in Figure 8.4.

Table 8.5 Responses to Social Network Question

Respondent Number	Respondent Name	Contact 1	Frequency	Value	Contact 2	Frequency	Value	Contact 3	Frequency	Value
1	Joe	Linda.	4	4	Brian	2	3	Val	1	1
2	Linda	Val	4	5	Jack	3	2	Joe	2	2
3	Yin	Jack	2	4	Li	4	2	Martin	1	2
4	Ahmed	May	3	3	Joe	2	2	David	3	2
5	Hussain	Ahmed	2	3	Joe	2	2	Val	2	2
6	Brain	Joe	3	5	David	3	2	July	3	1
7	July	Val	4	4	Li	2	1	Martin	2	4
8	John	Joe	3	5	Alison	3	3	Robert	2	2
9	Robert	John	2	5	Val	2	2			
10	Val	Joe	3	4						

SOURCE: Authors' own.

Table 8.6 Contact Matrix

		Joe	Linda	Yin	Ahmed	Hussain	Brain	July	John	Robert	Val	Jack	Martin	Li
Respondent	Joe	0	16	0	0	0	6	0	0	0	1	0	0	0
	Linda	4	0	0	0	0	0	0	0	0	20	6	0	0
	Yin	0	0	0	0	0	0	0	0	0	0	8	2	8
	Ahmed	4	6	6	0	0	0	0	0	0	0	0	0	0
	Hussain	4	0	0	6	0	0	0	0	0	4	0	0	0
	Brain	15	0	0	6	0	0	0	0	0	4	0	0	0
	July	16	0	3	0	0	0	0	0	0	0	0	6	0
	John	15	9	0	0	0	0	0	0	0	0	0	8	0
	Robert	0	0	0	0	4	0	10	0	0	4	0	0	0
	Val	12	0	0	0	0	0	0	0	0	0	0	0	0
	Jack	0	0	0	0	0	0	0	0	0	0	0	0	0
	Martin	0	0	0	0	0	0	0	0	0	0	0	0	0
	Li	0	0	0	0	0	0	0	0	0	0	0	0	0

SOURCE: Authors' own.

The size of the circles indicates how central or important someone is in the network and the thickness of the connecting line indicates the strength of the tie. Li is not that well connected and is said to be a pendant. Joe also acts as a bridge between fairly powerful actors of July, Ahmed, Linda and Val. From Figure 8.1 Joe is said to be very central, many cite him—he is in a position of power to influence others. A measure of assessing the degree of connectedness is centrality—this can be split into two parts: in-degree centrality, which is the weighted sum (by strength) of citations received and out-degree centrality the weighted sum of citations an actor gives. In-degree centrality can be thought of as a measure of power, whereas out-degree centrality can be thought of as an indication of deference. The centrality scores are presented in Table 8.7.

Note that John has high out-degree, which means he is asking questions, seeking information and consulting, while Joe has high in-degree power—he is being perceived as a source of knowledge. Also note that by questioning 10 people we have found out information on thirteen people. This does point to some ethical considerations. Usually to take part in a survey informed consent is sought, but how could have Jack, Martin or Li have given this—they were not revealed before the survey. For a tutorial paper on this see Pow et al. (2012).

Table 8.7 Centrality Scores for Discussion of Mobile Telephone Contracts

	Centrality	
Actor	Out Degree	In Degree
John	32	0
Linda	30	31
July	25	10
Brain	23	6
Joe	23	70
Yin	18	9
Robert	18	0
Ahmed	16	12
Hussain	14	4
Val	12	33
Jack	0	14
Martin	0	16
Li	0	8

SOURCE: Authors' own.

8.11 EXERCISE

Exercise 1

Design a short questionnaire to determine the factors that influence the frequency of people shopping at a particular supermarket, allowing analysis by gender, age and socio-economic grouping.

8.12 REFERENCES

B. Blumberg, D.R. Cooper and P.S. Schindler, *Business Research Methods* (Boston, MA: McGraw-Hill/Irwin, 2011).

S.P. Borgatti, M.G. Everett and L.C. Freeman, *Ucinet for Windows: Software for Social Network Analysis* (Harvard, MA: Analytic Technologies, 2002).

L.B. Bourque and E.P. Fielder, *How to Conduct Self-Administered and Mail Surveys* (London: SAGE Publications, 1995).

A. Bryman and E. Bell, *Business Research Methods* (Oxford: Oxford University Press, 2011).

J.R. Carpenter, M.G. Kenward and S. Vansteelandt, 'A Comparison of Multiple Imputation and Inverse Probability Weighting for Analyses with Missing Data', *Journal of the Royal Statistical Society, Series A*, 169, no. (3, July) (2006): 571–584.

R.L. Cross and A. Parker, 'The Hidden Power of Social Networks: Understanding How Work Really Gets Done in Organizations', *Harvard Business Review* (New York, NY, 2004).

D. Dillman, *Email and Internet Surveys: The Tailored Design Method* (Chichester: Wiley, 1999).

D. Futrell, 'Ten Reasons Why Surveys Fail', *Quality Progress* 27, no. 4 (1994): 65–69.

K. Gayen and R. Raeside, 'Social Networks and Contraception Practice of Women in Rural Bangladesh', *Social Science & Medicine* 71 (2010): 1584–1592.

P. Ghauri and K. Gronhaug, *Research Methods in Business Studies: A Practical Guide* (London: Prentice-Hall, 2002).

R.J.A. Little and D.B. Rubin, *Statistical Analysis with Missing Data* (Second edition) (New York: John Wiley, 2002).

J. Pow, K. Gayen, L. Elliot and R. Raeside, 'Understanding Complex Interactions Using Social Network Analysis', *Journal of Clinical Nursing* 21 (2012): 2772–2779.

J. Scott, *Social Network Analysis: A Handbook* (London: SAGE Publications, 2005).

T.E. Valente, *Social Networks and Health, Models, Methods and Applications* (Oxford: Oxford University Press, 2010).

8.13 WEBSITES

http://www.analytictech.com/
http://www.neighbourhoodcentre.org.uk/bank/bank.htm
http://www.statistics.gov.uk/harmony
http://www.questionpro.com
http://www.spss.com/spssmr
http://www.surveymonkey.com/

# CHAPTER 9

Interviews and Focus Groups

9.1 INTRODUCTION

A particular characteristic of business and management research is that knowledge of how things are, why they happen and what the intentions are is held by people. Consequently to obtain research data, in business and management, talking to people is important. How to do this is the subject of this chapter. First we begin by outlining the different purposes of interviews and who would be involved. Then some practical guidance is given. The chapter concludes by giving an overview of focus groups.

9.2 WHY DO INTERVIEWS?

As has been stressed, formulation is critical to successful research. Part of formulation in business and management should involve talking with the relevant stakeholders. These could be managers, employees or customers and one engages in general discussion about the chosen areas with the purpose of understanding what is important to the stakeholders, identify who are the important actors and how to access them. Such interviews are *exploratory interviews,* which should not probe into depth but should be general in nature; from this the scope and remit of your research get defined. Here you are obtaining an understanding of the system in which you will be researching and as such you are not answering research questions but figuring out how you will answer these questions.

If you follow a survey approach then before finalising the survey you should engage in *design interviews*. These are with the relevant stakeholders and targets of your research, perhaps employees in a motivation survey or customers in a customer satisfaction survey. These tend to be fairly short interviews, less than 20 minutes, and the aim is to ensure that important areas are covered by the survey, to test out some of the questions and to understand how best to administer the survey. This last point is important, as how to achieve the highest possible and most accurate response rate is vital, as is the need to ensure that disruption is minimised. Sometimes one can get some of the subjects of the research to complete a questionnaire, which you design in an interview format. Doing this will give you a wealth of valuable information as to the understandability of questions, what causes difficultly and how these questions could be re-designed to be more user friendly. Also and very importantly, this process helps identify any questions which may have been missed.

If you follow a more qualitative approach, then you may well engage in *in-depth research interviews*. These tend to last around one hour and probe behind the straightforward questions. These yield a vast amount of rich information. Often a semi-structured approach is taken; in this you produce a 'road map' of questions, which guide you through the interview. A question is asked and then you respond with more questions to the reply. For example, suppose you were interviewing customers about their perception of a clothes' fashion brand, then the outline of the semi-structured interview might be:

Clothing Brand Interviews

1. List three brands that you identify with.
2. For each brand tell me why you associate with them.
    Supplementary question 1: so based on these questions what type of person do you think you are?
3. Are you a customer of these brands?
    Supplementary question 2: for those brands you are a customer of in five years time do you think you will still be a customer of these brands.
    Supplementary question 3: for those brands which you are not a customer of—why not?
4. (show some advertising material) Ask to discuss what the advertising means to the individual.
    Supplementary question 4: ask them to try and explain why they have attached these meanings.
5. Give a list of emotional words such as anger, upset, romantic, loyal, happy, contentment, sexy, hostility, etc., and ask on a scale of 1–5 to what degree these are associated with the brand.
    Supplementary question 5: ask for some phrases to describe the brands and what they mean to the individual. This is with a view to collecting verbatim quotes.
6. Ask how frequently and how much they spend on the brand.
    Supplementary question 6: ask how satisfied they were with their purchases.
    Supplementary question 7: ask how this compares with other brands.
7. Ask them to describe the features of a brand they feel hostile to.
8. End with recording demographic characteristics of the respondent.

Often these cannot be predicted in advance, so you have to pay attention to the replies and 'think on your feet'. Successful in-depth interviews are difficult and require skill. To be effective, you must prepare and practise; more will be discussed on this in the next section.

People may be interviewed and re-interviewed over time in order to understand how their perceptions and attitudes change with time. Theses are called *longitudinal interviews*. This is a very powerful research methodology and can give tremendous insights. However, because this takes a long time—often many years—they tend not to be appropriate for undergraduate or Masters research. There is, however, an exception that lends itself to longitudinal study that is within the remit of an undergraduate or Masters study—this is the before and after interview. This is where interviews are conducted before and after an event in order to detect an effect. For example, this might be interviews about knowledge of, say, health and safety regulations before and after training or before and after acquisition by another company. In these before and after studies, it is important to detect and control for events external to the research area. For instance, has the wider economy changed, has the nature of competition changed or has the personal circumstances of the interviewee changed?

Finally there are *validation interviews*. The purpose of these is to try and determine if there has been a proper and reliable interpretation of collected data. Typically some of the research findings are presented to those who have been interviewed or surveyed as part of the main research to ascertain the degree of concordance with the research findings. These are *in-sample interviews*. One can also interview people who have not been previously surveyed or interviewed; these are called *out-of-sample interviews*. Feeding back findings to those who were not previously part of the study is a powerful way to ascertain the degree of generalisability to the results. To engage in this is highly recommended.

9.3 GENERAL GUIDELINES FOR INTERVIEWING

Choosing the sample: Those questioned should be representative of those whom you are attempting to make inferences about. They should also be the holders of the information you need to answer your research questions. Although the sample size should be determined statistically as interviews take a long time for undergraduate and postgraduate interview sample sizes tend to small (20–30 seems a good number). Of course, in doing case studies you may be dictated by size of the population. For example in doing a study of a human resource practice, there is a tendency to interview the HR staff. There may only be six staff and one might be on holiday!

Interviewing takes a long time and hence should be carefully planned. The actual interview rarely takes longer than one hour and rarely takes less than 20 minutes. The time element really comes from contacting people and persuading them to take part, travelling to meet them and after the interview analysing the content of the interview (this is the subject of a later chapter).

Always contact the subjects before the interview and organise when and where the interview will take place. This contact may be via a gatekeeper. A gatekeeper is someone who controls access to the research site. In a company you may be wishing to interview staff in a particular department—the gatekeeper in this situation could be the department head. Arrive early and be professional and courteous. Make sure you know exactly where to go. There have been many cases of people going to the wrong site or even the wrong company.

Always *pre-test* the interview schedule regardless of how structured or otherwise the interview is, how it is conducted (face-to-face or telephone) or how open or closed it is. Pre-testing the interview schedule on a small scale helps reveal and correct errors or problems. Several key questions should be asked at the pre-test stage:

- How comprehensive is the list of questions?
- Is the language appropriate?
- Are there other problems such as double meaning or multiple issues in one question?
- Does the interview schedule as developed help motivate respondents?

Try to pre-test not just on those likely to be participating in the study but give the interview schedule to friends and colleagues, for example, for comment.

At the beginning of the interview, explain what the interview is about and its purpose. Ask general simple questions first to put your subject at their ease. Often subjects have concerns about why they have been selected to participate in the study, so recognise this and explain why they have been selected and what you wish to achieve. Another concern that subjects have, especially in the workplace, is fears about anonymity and how the data are to be used. Often what you will be researching will not be confidential, but if the subjects have concerns that their comments may be reported to superiors, then this must be respected and you must keep the interview confidential and take steps to prevent the subject from being identified.

Then go into the more in-depth questions, keeping any sensitive ones to the end. Give flags that the interview is coming to an end, perhaps by saying "I have only a few more questions." Always end by asking if the subject would like to add anything. You can also ask the subject for suggestions on how to improve. Remember that things happen, perhaps the subject is called away, so try and ensure the really important questions are asked early on. Also remember that the subject is helping you, you are an inconvenience taking up time and intruding into their personal beliefs. So be polite even if you do not like or agree with what they say. For the subject the interview may be expensive—the hourly wage of a senior manager may be very high and to their company an hour of their time may be valued at five times their paid wage. So do not ramble on, keep to a strict time that you set at the beginning.

In conducting the interview the subject should do most of the talking—at least 80 per cent of the talk should be by the subject. Do not lead the subject, but similarly be firm—do not let the subject deviate from the area you are asking about. For important questions repeat in your words a summary of what you think the subject said to verify and check your interpretation. Perhaps the best skill of the interviewer is the ability to listen. This you should practise. Indeed before doing any interview have some practice by doing mock interviews on your friends. You should also actively show interest by using statements like 'that is very interesting' and smile and nod your head. Probe important questions and reflect the answers to these to the subject. So remember to listen actively, probe and reflect (LAPAR).

Try to be as consistent as possible when conducting the interviews on different subjects. Ask questions in the same order, have common questions across subject so that you can compare and contrast their answers and follow similar protocols about starting and ending. However, you will improve your interview skills the more you do so; if you find better ways of doing the interviews then by all means make changes. In free-flowing interviews, you should use checklists to help you to remember all the areas you wish to ask about. An example checklist is illustrated in Figure 9.1.

It is very important not just to listen to the questions and note the answers but also to observe body language, that is, the gesturers and tone used by the subject. Note what makes them excited, annoyed, happy, etc.—are they interested, enthusiastic, committed or bored. To do this you should split the paper you use to record answers into two-thirds for the answers and one-third for the observations. If you are permitted, it is a good idea

Figure 9.1 Example of an Interview Checklist

Checklists To aid memory and to allow free flow e.g., interview on training	

• Current training
  – Level/type
  – Why?
  – Where?
  – Benefits?
  – Meaning/Importance
• Future
  – What training would you like?
• Why?
	• Constraints
	  – Work/Home—time
	  – Money
	  – Supervisor's attitude
	  – Ability/qualifications
	  – Interest
	  – Reasons
	  – Past training
	• Personality
	• Career aspirations

SOURCE: Authors' own.

to use a tape or digital recorder to record the interview; this allows you to concentrate more on the interview and making observations. However, not all people agree to being recorded and there may be some validity issues; particularly in regard to sensitive issues, people may not be as forthcoming when they know their words are being recorded. Even if using a recorder you should still make notes of the answers in case the recording equipment fails and also in the notes you can underline or double or triple underline answers the subject feels strongly about.

In doing interviews it is good practice, if you can, to have a friend or colleague with you. This allows you to concentrate on the interview and they can make and note observations. Having someone with you also helps to build your confidence and ensure your safety. Be mindful about interviewing strangers; always do the interviews in public places or workplaces. Avoid interviewing strangers in their house or in your house. Always take an attitude of safety first. This may well be important if you visit workplaces and in some workplaces you may need to undergo a short course on workplace safety and wear protective clothing.

9.4 BIAS AND ERRORS

There are many potential sources of error and bias. The first and probably the most obvious is the simple misunderstanding of the question or the answer. Try including some related questions for cross-checking purposes. Another fairly obvious source of error is interviewer bias. Prejudice is not confined to the interviewer of course, it can be exhibited by the respondent as well. Even with goodwill and honesty a respondent may have memory failure and not realise that the response is untrue. Sometimes interviewees give responses that they think will please the interviewer. In addition to all of these are problems associated with the selection and sample.

The best protection against errors is an understanding of the various types. Two of the most important tactics in minimising bias are achieving rapport and trust with the respondent. Other tips for minimising bias and errors are:

- Getting to know the interviewees and their 'social context'
- Motivating individuals to respond (by stressing the importance of the work and their contribution to the study)
- Using appropriate questioning techniques such as funnelling questions, asking unbiased questions, clarifying issues and assisting individuals via probing. Numerous types of probes exist, including basic probes such as
    - Repeating the question if the interviewee strays
    - Explanatory probes (e.g., 'what did you mean by that')

- Focused probe (e.g., 'what kind of …?')
- Giving ideas (e.g., 'have you tried …?', 'have you thought about …?')
- Reflecting probes (e.g., 'what you seem to be saying is …')

where possible always try to verify responses from interviews with data from other sources.

9.5 TELEPHONE INTERVIEWS

These have some particular advantages (over face-to-face interviews) such as ease of geographic coverage, the possibility of doing more interviews in the time available and costs are usually lower. However, significant drawbacks exist and these include the sense of impersonality especially if you have not met before (which may make rapport difficult to achieve), the lack of visual contact (cannot read non-verbal cues) and there builds a feeling of time pressure leading to a tendency to rattle through the interview. Some consider that telephoning for an interview is intrusive for the party receiving the call and the interviewer's awareness of that to be a major problem. In addition, students are often nervous about using the telephone for interviewing and transmit their nervousness to the respondent. Generally telephone interviews seem more prone to interruptions and early termination. All of these can affect the quality of the data obtained.

Tips for overcoming these drawbacks are as follows:

- Use voice cues to compensate for the lack of visual contact (e.g., 'yes', 'good', 'have a think about that for a minute', etc.).
- Listen sensitively and do not talk too much.
- Remember the importance of your tone and the need to project warmth and friendliness.
- Remember to write and take notes as well as listen (which is easier than in the face-to-face interviews).
- Sum up important points from time to time.
- Arrange in advance a mutually convenient time for the interview.
- Jot down what you wish to say; bring all necessary materials/papers to the phone.
- Do not be afraid of silence (you can always check on what is going on by asking such questions as 'did you understand my last point?', 'do you need to think about that?' etc.).

9.6 GROUP/FOCUS GROUP INTERVIEWS

Group interviews can lead to efficiencies in that responses from several subjects can be elicited simultaneously, thus saving a considerable amount of time. We have found them particularly useful for validating findings. Perhaps when a survey is conducted it is good practice to get a group together who completed the survey and discuss your findings and inferences with them. This helps you to identify if you have made the correct interpretation and to assess the representativeness of the survey. Other group interviews could be held with those who did not take part in the original survey.

In group interviews the groups should be composed of as homogeneous people as possible, with similar grades, experience and ages as we are try to identify the group view. The view of one group could then be compared with that of other groups.

Morgan (1988) gives good advice on the running of focus groups, which as the name suggests focuses on specific issues and is a favourite tool of market researchers. The optimal size for a focus group is between eight and twelve subjects and it is important that all participate and no individual dominate the discussions. The role of the researcher becomes one of facilitation to guide the group to discuss on the relevant issues, keep the group focussed on these issues rather than deviating and ensuring that all participate. Although facilitation skills are important, so are those of observation and listening. To successfully use focus groups, at least two researchers are needed. One facilitates and asks the questions and the other notes responses and observations. Clearly, recoding proceedings—either by tape recorder or preferably on video—is a good idea.

Focus groups take considerable planning and this must be carefully done. It takes time to contact the subjects, organise and book the venue and plan the questions. Often the subjects are sent material in advance so that they too can prepare for the group interview. Each group interview session will last at least two hours and we advise not to exceed three hours. Organising and documenting all the findings take a considerable amount of time and each group interview should be documented before embarking on the next set.

There are some problems which can arise and need to be guarded against. One is that subjects do not participate, perhaps because of shyness. To overcome this, giving the material well in advance and explaining the background at the beginning of the session can help. Then systematically ask each individual for their general views. This helps to break the ice. Another problem is that one or two people dominate the discussion. Here the facilitator should politely but firming request that they limit their opinions and let others have their say. Also 'group think' can occur, where subjects fall into line with the group view, which can be established by the first to speak. Subjects fear embarrassment from contradicting the group. A way to minimise this is to ensure different subjects begin the debate and a debate with them as to what could be weak parts in their view. An advantage of group interviews is that differences of opinion can be debated.

Undergraduate and postgraduate students rarely undertake group/focus group interviews but do not let this put you off. This type of interview offers you the possibility of collecting a lot of information very quickly and fits well with the typical dissertation period of three to four months.

9.7 REFERENCE

D.L. Morgan, *Focus Groups as Qualitative Research* (Newbury Park, CA: SAGE Publications, 1988).

CHAPTER 10

Qualitative Data Analysis

10.1 INTRODUCTION

Qualitative research poses quite a challenge to the researcher in terms of how to reduce what may feel like an overwhelming amount of data gathered from in-depth interviews, observations, written documentation and so on. There are a number of approaches to help you do this. Some rely on the use of sophisticated computer packages, and others on more traditional manual techniques. For the type of research typically involved in an undergraduate or Masters level study the volume of data collected tends to be manageable by traditional means. We would only use computer packages if you have a lot of interview data (say more than 40 interviews) or textual content to analyse. Packages such as NVIVO or SPSS text analyser are useful. If you are going to analyse qualitative data using software then we advise you seek some specialised instruction and begin by reading the section in Silverman's (2005) book on qualitative analysis. This book covers the concepts of traditional analysis as well.

The aims of analysis are seven-fold:

1. Detect patterns
2. Identify deviants and oddities
3. Compare to theory—detection of conformance (if the scientific method is used)
4. Identify groups—classification
5. Compare and contrast groups
6. Construct a model
7. Test the model—validation

Thus there is an exploration phase (aims 1 and 2), a classification phase (aims 3 and 4), a drawing conclusions' phase (aims 3 and 5), a representation phase (aim 6) and a testing phase (aim 7).

10.2 PREPARATION

To start the analysis the qualitative data has to be prepared—in the scientific method this is inevitably reductionist. All the observations, inflections in tone and the way words were combined can't be captured. But you try to keep as much as possible. Many seem to think interviews, etc., should be fully transcribed, but although this may make you feel good by doing a lot of work we think it is rather pointless (especially if the interview is stored on tape or preferably on some electronic medium). Rather, we recommend that for observation and interviews, notes should be constructed. These should be in two forms as illustrated in Table 10.1.

Once this is done then a useful approach is to write summarised information in the form of statements onto 'post-its'. This should be at the lowest level of analysis and each 'post-it' should contain only one point—do not use 'and' or 'or'. Also put verbatim quotes onto 'post-its'. 'Post-its' of different colours, sizes and shapes can be used to represent different times, different grades of employment, different divisions, companies, etc. Then these 'post-its' can be placed on a blank wall and moved around into groups representing themes and placed into clusters by statements they have in common. Classifications and order will emerge from the initial chaos. Once this exercise is complete, which will take several days even for a modest amount of interviews (say

Table 10.1 Reducing Interview and Observational Data

Type of Data	Transcription	Other Information
Interviews	Main points and key or interesting verbatim quotes. Type points of information in black, attitudes, feelings, view, etc., in green and verbatim quotes in blue (or any other colours you choose)	Create a separate but linked interview code number list of observations about: a. Demographic/employment characteristics b. The interviewee—body language, tone, when they got emotional, etc. c. Environment of the interview—disruption, temperature, clutter, etc. d. External environment—news reports which might influence the interview, market conditions, other issues influencing the company that day e. Any other relevant information
Observations	a. Description of the physical environment b. Description of individuals' physical characteristics c. Description of individuals' behaviour d. Description of interactions	a. External environment—news reports which might influence the interview, market conditions, other issues influencing the company that day b. Any other relevant information

SOURCE: Authors' own.

around 20), information can be summarised by theme and frequencies produced. Typical cross-tabulations can be formed to see how understanding, views and attitudes of people by different grades, gender, experience, etc., compare.

To illustrate this, taped interviews were taken of 12 managers in a major brewing company to determine their level of understanding, involvement and attitude to sales forecasting in their company. From the notes of the interviews over 400 'post-it' notes were made and these were tagged to a wall of a room in a very rough way. They were listed under the headings of: 'Trade Marketing and Planning', 'Trading', 'Customer Service and Operations' and 'Finance'. Organising the statements into this taxonomy took several days. It is unwise to do such classifications quickly as unless periods to reflect on the classification are built in, sub optimal segmentation will result.

Within the headings the responses were then ordered by theme and division to give further categorisation and to allow Tables 10.2 and 10.3 to be constructed.

Table 10.2 Frequency Count of Responses by Item and Respondent

| | | Trade Marketing and Planning | | | |
		Forecasting Manager	Business Palnning Manager	Promotional Analysis Manager	Total
Process	Purpose	1	1	1	3
	Horizon/Plans	0	0	1	1
	Data sources	3	1	1	5
	Production	16	3	3	22
	Organisation (flow)	0	2	1	3
	Organisation (review)	1	0	3	4
	Supply chain	1	2	1	4
	Total	22	9	11	42
Usefulness	Perception	5	0	1	6
	Bias	1	4	1	6
	Accuracy	9	7	1	17
	Total	15	11	3	29
Forecasting environment	Culture/Change	1	4	8	13
	Promotions	2	1	1	4
	Customers	5	4	1	10
	Business environment	3	4	8	15
	Total	11	13	18	42
Total		48	33	32	113

(Table 10.2 contd)

(Table 10.2 contd)

		Trading				
		Account Director (Wholesale)	Business Account Executive (Grocery)	Account Director (Grocery)	Business Account Manager (Convenience)	Total
Process	Purpose	3	2	1	1	7
	Horizon/Plans	4	1	1	0	6
	Data sources	5	2	1	3	11
	Production	6	3	7	2	18
	Organisation (flow)	2	0	1	1	4
	Organisation (review)	3	4	1	4	12
	Supply chain	0	1	3	1	5
	Total	23	13	15	12	63
Usefulness	Perception	1	2	3	2	8
	Bias	2	4	6	3	15
	Accuracy	0	1	1	3	5
	Total	3	7	10	8	28
Forecasting environment	Culture/Change	1	0	1	1	3
	Promotions	5	0	6	6	17
	Customers	10	2	5	1	18
	Business environment	5	2	4	3	14
	Total	21	4	16	11	52
Total		47	24	41	31	143

		Customer Service and Operations			Finance		Total
		Demand Planner	Demand Plan Manager	Total	Financial Planning Manager	Total	
Process	Purpose	4	5	9	3	3	22
	Horizon/Plans	2	0	2	4	4	13
	Data sources	1	0	1	2	2	19
	Production	9	3	12	3	3	55
	Organisation (flow)	4	2	6	3	3	16
	Organisation (review)	3	1	4	3	3	23
	Supply chain	10	1	11	0	0	20
	Total	33	12	45	18	18	168

(Table 10.2 contd)

(Table 10.2 contd)

		Customer Service and Operations			Finance		Total
		Demand Planner	Demand Plan Manager	Total	Financial Planning Manager	Total	
Usefulness	Perception	2	0	2	1	1	17
	Bias	1	2	3	1	1	25
	Accuracy	14	6	20	1	1	43
	Total	17	8	25	3	3	85
Forecasting environment	Culture/Change	2	8	10	0	0	26
	Promotions	2	0	2	0	0	23
	Customers	0	1	1	0	0	29
	Business environment	5	4	9	3	3	41
	Total	9	13	22	3	3	119
Total		59	33	92	24	24	372

SOURCE: Canduela (2006).

Table 10.3 Responses by Department (Frequency Count)

Responses by Department	Trade Marketing and Planning	Trading	Customer Service and Operations	Finance
Process	42	63	45	18
Usefulness	29	28	25	3
Forecasting environment	42	52	22	3
Total	113	143	92	24

SOURCE: Canduela (2006).

It was concluded that:

- All departments except Finance were similar in terms of the proportion of responses associated with each forecasting element.
- The highest response by individuals was that of Customer Services and Operations, showing that this department is the one being most affected by issues related to the forecasting function and therefore the one that has the most points to discuss.

- The most important element for all four departments was the Process, especially for the Finance department.

A model of forecasting was then formed and this is displayed in Figure 10.1.

In essence this outlines qualitative analysis if the scientific method is used and suffers from the criticism of being reductionist and 'forcing quantification'. The approach outlined above was rather ad hoc and a more structured approach to analysis is called the 'Framework Approach', which was developed by Ritchie and Spencer (1994), social policy researchers.

The approach comprises five stages, each of which is interrelated:

- Familiarisation of the data
- Creating a thematic framework
- Indexing
- Charting
- Mapping and interpretation

These steps are now elaborated on.

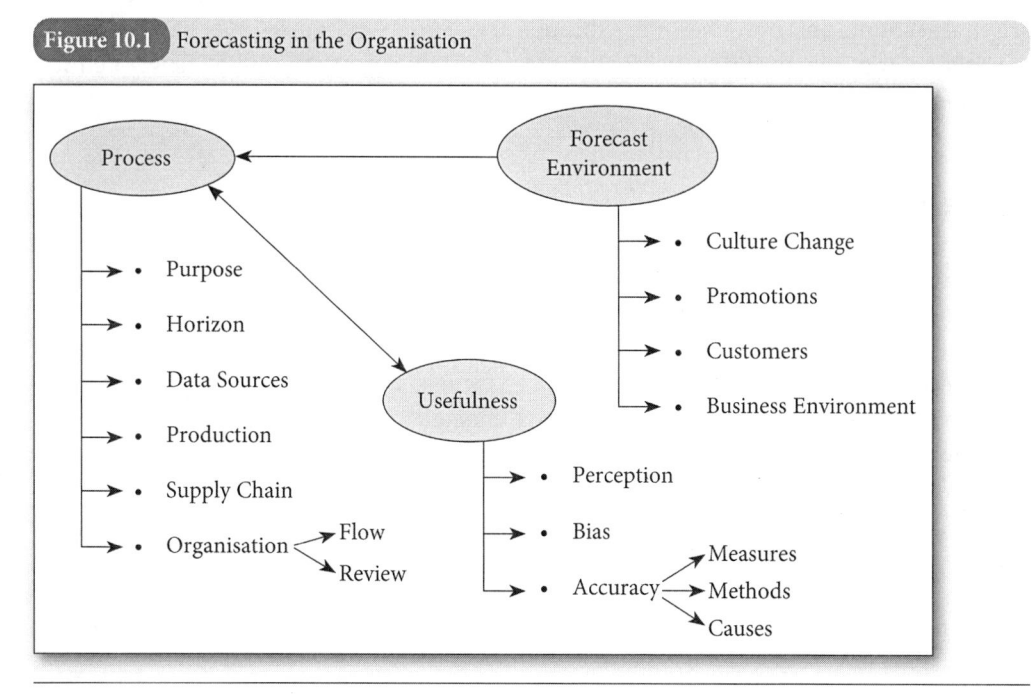

Figure 10.1 Forecasting in the Organisation

```
                           Forecast
      Process  ◄────────  Environment
         │                     │
         │                     ├──► • Culture Change
         ├──► • Purpose        ├──► • Promotions
         ├──► • Horizon        ├──► • Customers
         ├──► • Data Sources   └──► • Business Environment
         ├──► • Production         
         ├──► • Supply Chain   Usefulness
         └──► • Organisation ◄ Flow
                              ◄ Review
                                 │
                                 ├──► • Perception
                                 ├──► • Bias
                                 │              ┌► Measures
                                 └──► • Accuracy─► Methods
                                                └► Causes
```

SOURCE: Canduela (2006).

Familiarisation of the Data

This involves immersion in the data in order to gain an overview of the depth and diversity of the material, and the identification of recurring themes and issues.

Miles and Huberman (1994) describe a number of methods useful for early familiarisation and analysis of research material. For example, writing up a contact summary sheet as soon as transcripts have been typed up provides a useful reflective overview of what went on during the interview or discussion, much of which would have been lost if summarised later.

These sheets also help direct planning for subsequent interviews and the development of topic guides and generation of categories, which have become more sophisticated as new territories unfold. They provide an invaluable source of keeping the researcher on track at a time when he/she could become buried in a large volume of detail.

Creating a Thematic Framework

Having identified a number of key themes at the familiarisation stage, these are examined in detail with a view to setting up a thematic framework or index. This requires careful reading through of transcripts in full to ensure that any index is grounded in the original accounts and observations. Although some index categories may be identical to the original topic areas and questions covered at interview, others will be newly defined from the emergent themes.

Coding and Indexing of Data

The next stage involves applying codes to the data. Codes can be numerical (see example provided) or alphabetical. Miles and Huberman recommend the use of alphabetical codes because it helps ensure that the researchers keep close to their data. All your material is read through and coded alongside the margin of the text. Single paragraphs often contain a number of different themes and are coded accordingly.

The data should be well moulded to the codes that represent them. This is not a routine exercise and is very time consuming, but nevertheless can provide a useful mechanism for reviewing judgements made about the meaning and significance of the data.

Charting

Distilled data summaries can now be charted according to 'coded' text. This allows the analyst to build up a picture of the data as a whole.

Mapping and Interpretation

The final stage in this process of analysing qualitative data is about interpreting the data and making assertions. The analyst reviews the charts and research notes, compares and contrasts perceptions, accounts or experiences and searches for patterns and connections that will help explain the phenomenon under study.

This process will be shaped by the key objectives and research questions guiding the study, and can be clustered under the following headings identified by Ritchie and Spencer (1994).

- Defining concepts
- Mapping range and nature of phenomenon
- Creating typologies
- Finding associations
- Providing explanations
- Developing strategies

Wolcott (1990) provides some useful advice regarding the write-up of qualitative research. For a comprehensive guide to the analysis of qualitative material, see the International Institute for Qualitative Methodology at: http://www.ualberta.ca/~iiqm/.

10.3 CONTENT ANALYSIS

This is a popular approach to the analysis of qualitative information. Key phrases or words are counted and the frequencies analysed. The selection of these depends on, for example, the particular hypotheses to be tested. This method may be useful in allowing the researcher to present a picture of what the concepts are but it does not help in understanding why ideas or interpretations arose in the first place. If you engage in qualitative research it is likely that you will engage in some form of content analysis as content analysis is frequently used to analyse text, pictorial information, interviews and web pages.

As the name suggests its purpose is to describe the content of your respondents' comments systematically and classify the various meanings expressed in the material you have recorded. It isn't the only way in which you would analyse the data you have obtained, and you might find yourself presenting information in the form of a connected narrative (in a study following the case study method, for example), or by means of a series of verbatim quotations taken from the interviews. Another way is to 'tell a story'.

But all reporting of semi-structured interviews assumes that you present findings which are representative of what was said, and content analysis is a powerful means for familiarising yourself with what's there, quite apart from your readers. There are six main steps to follow in content analysis.

Identify the Unit of Analysis

Often this is obvious it might be the individual, or the company, but it may be more generic like country.

Choose a Set of Categories

These must be relevant to the issue being explored, mutually exclusive (so that a unit can only be placed under one category), exhaustive (covering all possibilities), and reliable (someone else repeating the analysis would categorise the unit in the same way). This is done in two ways: either because a theory or rationale prescribes them, or because your review of the material suggests them as useful.

It is also important that the categories that you construct are reliable. If you are using categories drawn from a pre-existing theory or rationale, then it's important that, before you use them to classify your data, you make a brief, explicit list of the defining characteristics of each category: that is, of the signs that you will be looking for in order to put each assertion into one category rather than the other. Reliability is a bigger problem in those situations in which you have no pre-existing categories in mind, and have to draw them up in the first place as a result of reading and perceiving the dominant themes in your interview transcripts. That's because there is no pre-existing list of defining characteristics, and because you have to invent these, as well as the categories themselves. In either case, the issue of reliability boils down to a simple question. Would someone else perceive the same categories as you did? The way to answer this question is to hand a photocopy of the uncoded transcripts to a colleague whom you ask to recognise categories for him- or herself. You would then compare the two category sets, your own and theirs, and argue over them until something more useful emerged and agreement on the defining characteristics was obtained. This would then be tested by seeing if you both agreed on the coding (assigning of comments to categories). You should consider involving one or more of your respondents in this role.

Coding

Read through the material, and, within each context unit, assign each assertion to one of the categories. There may be more than one assertion within a context unit.

Coding sheets are sometimes developed to assist with this process. For example content analysis could be used to analyse web pages of companies to ascertain the companies' attitude to corporate social responsibility (CSR).

The coding sheet could be:

Company name: ………………….
Web address: ……………………   Date: ……………..
Number of pages: ……………..   Number of pages with CSR statements: …
Percentage of content which is related to CSR statements: …….
CSR certification: …………………..   Date of certification: …………………
Areas of organisation covered by CSR: ………………………………
Percentage of organisation involved in CSR: ………………………
Examples of CRS statements: ………………………………..
Examples of CSR projects and dates: ……………………………………

If one was researching how images of culture and heritage are used in holiday advertising one could analyse the content of holiday advertisements in newspapers. The following coding sheet could be used.

Newspaper: ……………………..   Date: ……………………
Holiday adverts: …………………
Number of adverts which contain a culture/heritage message: ….
Culture/Heritage advert 1.
Size of advert: ……………….
Colours: ……………   Pictures: ……………
Images of culture/heritage: ……………………………………..
Percentage of advert that is devoted to culture/heritage: …………….
Percentage of images that are devoted to culture/heritage: …………….
Percentage of text that is devoted to culture/heritage: …………….
Judgemental mark out of 10 on appeal of culture/heritage message: ….
Judgemental mark out of 10 on accuracy of culture/heritage message: ….
Culture/Heritage advert 2.
Size of advert: ……………….
Colours: ……………   Pictures: ……………
Images of culture/heritage: ……………………………………..
Percentage of advert devoted to culture/heritage: …………….
Percentage of images that are devoted to culture/heritage: …………….
Percentage of text that is devoted to culture/heritage: …………….
Judgemental mark out of 10 on appeal of culture/heritage message: ….
Judgemental mark out of 10 on accuracy of culture/heritage message: ….

Tabulate the Material

Count the number of assertions under each category and present the material as a table.

Illustrate the Material

Present the categories and list the assertions under them: all, or a representative set.

For example from interviews of staff, their view of senior management on a broad scale of negative to positive was found. How this is related to their experience in the company can be found from a simple graphic shown in Figure 10.2.

The numbers represent the identity of each of the 10 members of staff who were interviewed. This may be interpreted that person 8 is new to the company and scores the management high—perhaps grateful for the job and being keen to fit in. However, staff 1, 2 and 9 have more experience and have become rather jaded. Staff 10, 6 and 4 have more positive views of management—perhaps these have been promoted or otherwise given recognition. Something to check in the interview notes. Then staff 5, 3 and 7 have a low opinion of management—perhaps these people with lots of experience have been passed over for promotion.

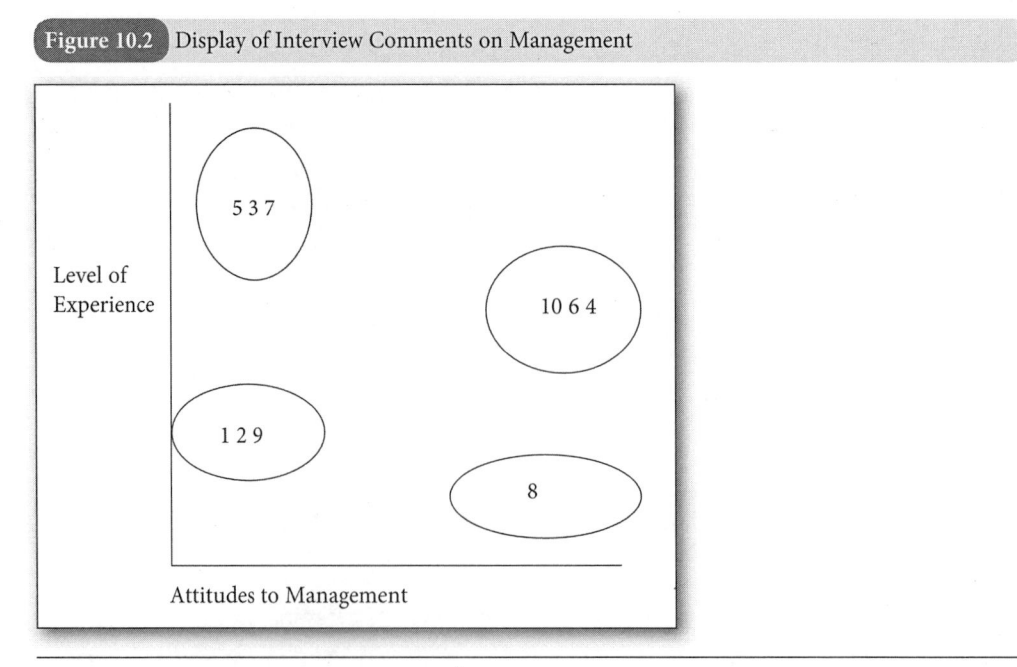

Figure 10.2 Display of Interview Comments on Management

SOURCE: Authors' own.

Force Field Analysis is another way of depicting qualitative data and helps to categorise positive and negative influences. For example, influences on future revenue of a company gleaned from interviews of members of the board of directors can be illustrated in the diagram shown in Figure 10.3.

Schematic diagrams are also useful for illustrating findings from interviews. From interviews of staff in company X as to what contributes to performance, one can summarise and illustrate ideas by a figure like the one illustrated in Figure 10.4.

There is a strong feeling that experience has a positive influence on performance, then the next strongest but some way behind is training of staff and there is a slight positive influence provided by incentives. Increased intensity of production is considered to have a negative influence. The strength of influence is shown by the thickness of the line.

Cause and Effect are diagrams frequently used in quality management to illustrate how factors may contribute to a target effect and help to order these into broad

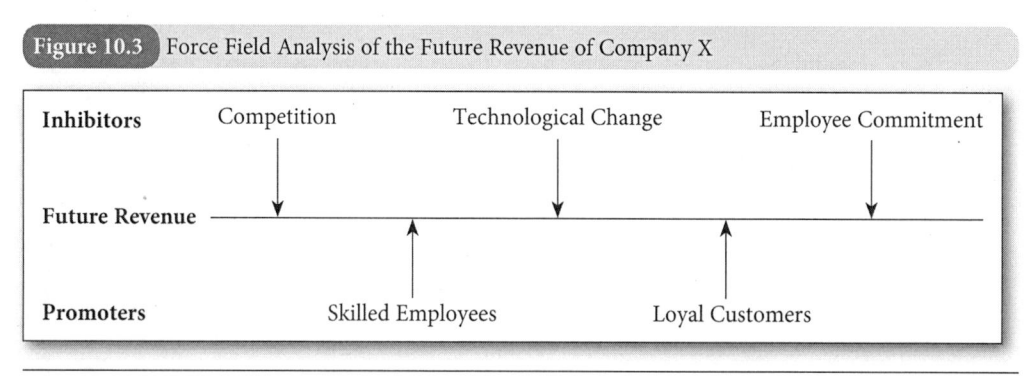

Figure 10.3 Force Field Analysis of the Future Revenue of Company X

SOURCE: Authors' own.

Figure 10.4 Schematic Diagram of Contributors to Performance for Company X

SOURCE: Authors' own.

Figure 10.5 Cause and Effect Diagram of Factors Leading to Absenteeism

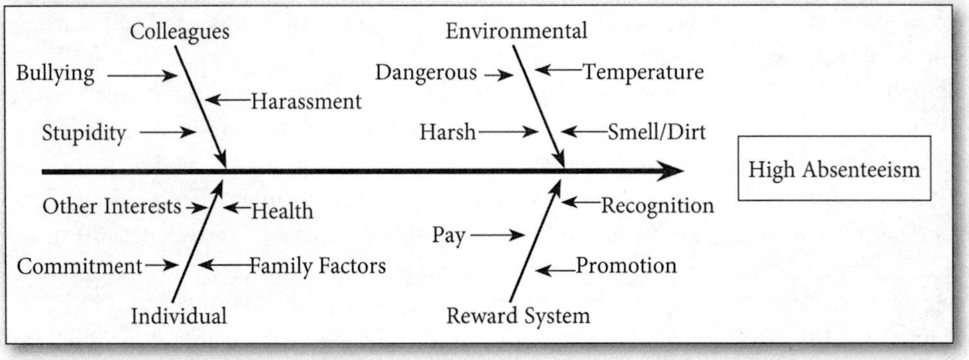

```
        Colleagues              Environmental
Bullying ──→         Dangerous →   ←Temperature
         ←─Harassment
Stupidity ──→        Harsh ──→    ←Smell/Dirt
─────────────────────────────────────────────→  High Absenteeism
Other Interests →/ ←Health       ←Recognition
                       Pay ──→
Commitment →/ ←Family Factors    ←Promotion
        Individual              Reward System
```

SOURCE: Authors' own.

categories. Consider a situation in which staff are interviewed as to what they think causes absenteeism in their organisation. The diagram displayed in Figure 10.5 might be produced.

In Figure 10.5 high absenteeism is the target effect and the main categories of contributing factors are colleagues, environment, individual and the reward system.

To study communication flows in a company, staff may be interviewed or observed as to who they mainly communicate with. Communication flows can be represented by a map called a sociogram. A sociogram generated from company X is illustrated in Figure 10.6.

The thickness of the connecting lines shows the strength of communication and the arrowheads show the principle direction of communication. This figure indicates that company X may have a number of problems—the manger seems to communicate principally with the supervisor and secretary, the engineer feels rather 'out of it'. The engineer also does not communicate well with store. If Worker C does not know what to do or does not follow health and safety rules, it may not be entirely their fault as they are isolates in the web of communication.

10.4 SUMMARISING

Tabulation is the usual way of presenting the information available in content-analysed data. If you want to make inferences from the numbers of utterances under each category, then a table to this effect, perhaps stratified according to your sampling design, will reveal relationships to you, and permit a variety of descriptive and analytic statistics to be carried out on the numbers. If you're less concerned with the numbers, then a list of

Figure 10.6 Communication Flows in Company X

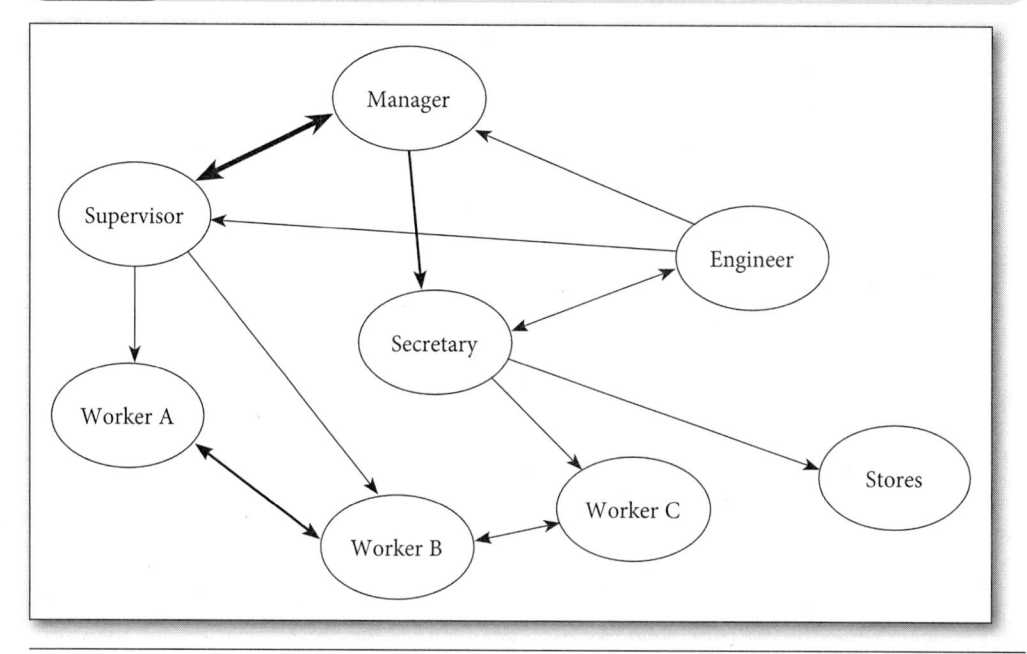

SOURCE: Authors' own.

the various comments, or representative examples, can be reported in a table in which the columns are the category headings that you've used. The following table presents the results of a content analysis of part of a semi-structured interview in which respondents were asked about their views of a new diversification plan.

Researcher's Notes

Sample: All 33 members of New Product Development department
Stratified: 17 in the Industrial Products division, 16 in the Consumer Products division
Respondents: 32 (17 and 15 + 1 unavailable on the day of interview)
Recording unit: What was said
Context unit: That part of the interview in which respondents discussed their views of the new diversification plan: whole of the reply treated as an entry under one category, regardless of number of utterances
Data: (A transcript of the 32 conversations with the relevant part of the interview highlighted, each coded with a number, 1–6, according to the categories below)

Categories: (Derived from consideration of the highlighted data, defining characteristics in italics):
Plan inappropriate in view of what competitors are doing
Plan unviable since the proposed divestments are inappropriate
Plan unviable since planned acquisitions don't match company policy
Plan viable but needs further development
No view either way
Miscellaneous

From the analysis of the interviews Table10.4 can be formed:

The majority of respondents (82 per cent) expressed unfavourable views about the plan, the most common being that competitors' activities were insufficiently taken into account. This is true in each of the Divisions in which the respondents are located. Only a small minority (1 person in each Division) felt the plan was viable, and that with reservations; one further person refused to commit him/herself. The miscellaneous comments concerned some of the personalities involved and does not add to the information expressed.

Categories were developed by looking at the different answers given in the transcripts of that part of the interview. As presented, the information is formally identical to what could have been obtained by a structured interview or questionnaire question (largely due to the context unit being set to equal to the respondent's entire comment in response to the question, regardless of how many sentences were involved), and many of the reporting conventions of the more structured techniques apply.

Table 10.4 Findings from Interviews on New Diversification Plan

	Industrial Products		Consumer Products	
	n	%	n	%
Plan inappropriate in view of what competitors are doing	8	47	7	44
Plan unviable since the proposed divestments are inappropriate	4	24	3	19
Plan unviable since planned acquisitions don't match company policy	3	18	3	19
Plan viable but needs further development	1	6	1	6
Miscellaneous	0	0	1	6
No view either way	1	6	0	0
No answer	0	0	1	6
Total	17	101	16	100

SOURCE: Authors' own.

This is not surprising. The answer categories for fully structured interviews and questionnaires are obtained through content analysis in the first place, whether the categories derive from some theory, or are developed on the spot following the above procedure, or are developed informally 'in one's head' on material obtained by conversational techniques.

10.5 GROUNDED THEORY

This recognises the fact that large amounts of non-standard data produced by qualitative studies make data analysis problematic. Unlike in quantitative analysis earlier where an external structure is imposed on the data, some qualitative researchers require that structure be derived from the data itself and requires the data to be systematically analysed to bring out themes or patterns.

In seeking to apply grounded theory to data transcripts it is usually best to start with a thorough familiarisation with the data and the framing of questions leading to reflection, which involves trying to make sense of the data. A critical appraisal of the data is necessary at this stage including an awareness of previous research, etc. Conceptualisation comes next, whereby a set of concepts or variables appear important in understanding what is going on. Concepts are now articulated as explanatory variables but the researcher should still be sceptical over their validity, reliability, etc. Cataloguing concepts comes next but whose language to use, researcher's or respondent's? Re-coding occurs once all references to particular concepts are known and we can go back to the original occurrences to see if there is a case for re-assigning these.

The next stage is linking, where all of the variables considered to be important can be linked together towards a more holistic theory. This process involves consideration of literature and existing models, etc., and relating this to the results. Finally, re-evaluation should take place in the light of peer scrutiny, etc.; perhaps more work is required, or perhaps a different emphasis should be placed on some aspects of interpretation, etc.

This approach is good for the development of new concepts and theories but it really requires a great deal of skill, experience and most importantly time. Often there are several iterations through observation and concept/theory forming. Although grounded theory has allowed many great insights to be made in business and management because of the need for high skill levels, training and large amounts of time, we do not recommend grounded theory as an approach to follow at undergraduate or Masters level. A more comparative critical discussion on qualitative inquiry and research designs can be found in Creswell's (1998) book.

10.6 REFERENCES

J. Canduela, 'Forecasting in Fast Moving Consumer Goods Organisations' (PhD thesis, Edinburgh Napier University, 2006).

J.W. Creswell, *Qualitative Inquiry and Research Design: Choosing among Five Traditions* (London: SAGE Publications, 1998).

M.B. Miles and A. M. Huberman, *Qualitative Data Analysis* (London: SAGE Publishing, 1994).

J. Ritchie and L. Spencer, *Analysing Qualitative Data* (London: Routledge, 1994).

D. Silverman, *Doing Qualitative Research* (Second edition) (London: SAGE Publishing, 2005).

CHAPTER 11

Descriptive Quantitative Analysis

11.1 INTRODUCTION

In recent years, there has been a move towards evidence-based decision-making. To do this effectively, business researchers must collect numerical data, store it and interrogate it. In this chapter, an overview of some statistical techniques that you could well find useful in your research is presented. The chapter begins with a review of basic statistical summaries: frequencies, bar charts and numerical measures and tables. It covers the use of *t*-tests to compare two groups as well as analysis of variance (ANOVA) to compare more than two groups. These are tests for continuous data. The chapter concludes with how to test for association between categorical forms of data.

In this chapter, examples of how to conduct the analysis in a spreadsheet such as in Microsoft Excel are given. However, Excel is not a very flexible package to analyse the data in, and a more user-friendly and popular statistical analysis software package is the Statistical Package for the Social Sciences (SPSS). You can find out more about SPSS by visiting the SPSS website on www.spss.com (see also Bryman and Crammer 2004). Excellent books to consult are *Doing Statistics for Business with EXCEL* by Pelosi and Sandifer (2002) and *Understanding Statistics Using SPSS* by A. Field (2009).

Consider an example in which a bank has collected data on the starting salary, current salary, number of years of education, age, gender, job category and number of years of employee service for each of its 474 employees. This data has been stored in the Excel file called employe.xls and is available on the web page http://www.sagepub.in/books/Book242499/samples (see Figure 11.1).

We have discussed types of variables and their measurements in the previous chapters. However, to generate data we must need sufficient knowledge about coding and recoding of variables. This job is a little tedious for primary sources of information but is relatively easy for secondary sources. For primary sources, we need sufficient editing

Figure 11.1 Bank Employment Dataset

[Screenshot of a Microsoft Excel spreadsheet titled "empl" showing the Bank Employment Dataset with columns: ID, GENDER, EDUC, JOBCAT, SALARY, SALBEGIN, JOBTIME, PREVEXP, MINORITY]

ID	GENDER	EDUC	JOBCAT	SALARY	SALBEGIN	JOBTIME	PREVEXP	MINORITY
1	m	15	3	$57,000	$27,000	98	144	0
2	m	16	1	$40,200	$18,750	98	36	0
3	f	12	1	$21,450	$12,000	98	381	0
4	f	8	1	$21,900	$13,200	98	190	0
5	m	15	1	$45,000	$21,000	98	138	0
6	m	15	1	$32,100	$13,500	98	67	0
7	m	15	1	$36,000	$18,750	98	114	0
8	f	12	1	$21,900	$9,750	98	0	0
9	f	15	1	$27,900	$12,750	98	115	0
10	f	12	1	$24,000	$13,500	98	244	0
11	f	16	1	$30,300	$16,500	98	143	0
12	m	8	1	$28,350	$12,000	98	26	1
13	m	15	1	$27,750	$14,250	98	34	1
14	f	15	1	$35,100	$16,800	98	137	1
15	m	12	1	$27,300	$13,500	97	66	0
16	m	12	1	$40,800	$15,000	97	24	0
17	m	15	1	$46,000	$14,250	97	48	0
18	m	16	3	$103,750	$27,510	97	70	0
19	m	12	1	$42,300	$14,250	97	103	0
20	f	12	1	$26,250	$11,550	97	48	0
21	f	16	1	$38,850	$15,000	97	17	0
22	m	12	1	$21,750	$12,750	97	315	1
23	f	15	1	$24,000	$11,100	97	75	1
24	f	12	1	$16,950	$9,000	97	124	1
25	f	15	1	$21,150	$9,000	97	171	1
26	m	15	1	$31,050	$12,600	96	14	0
27	m	19	3	$60,375	$27,480	96	96	0
28	m	15	1	$32,550	$14,250	96	43	0
29	m	19	3	$135,000	$79,980	96	199	0
30	m	15	1	$31,200	$14,250	96	54	0
31	m	12	1	$36,150	$14,250	96	83	0
32	m	19	3	$110,625	$45,000	96	120	0
33	m	15	1	$42,000	$15,000	96	68	0
34	m	19	3	$92,000	$39,990	96	175	0
35	m	17	3	$81,250	$30,000	96	18	0
36	f	8	1	$31,350	$11,250	96	52	0
37	m	12	1	$29,100	$13,500	96	113	1
38	m	15	1	$31,350	$15,000	96	49	1
39	m	16	1	$36,000	$15,000	96	46	1
40	f	15	1	$19,200	$9,000	96	23	1
41	f	12	1	$23,550	$11,550	96	52	1
42	m	15	1	$35,100	$16,500	95	90	0
43	m	12	1	$23,250	$14,250	95	46	0
44	m	8	1	$29,250	$14,250	95	50	0

SOURCE: Authors' own.

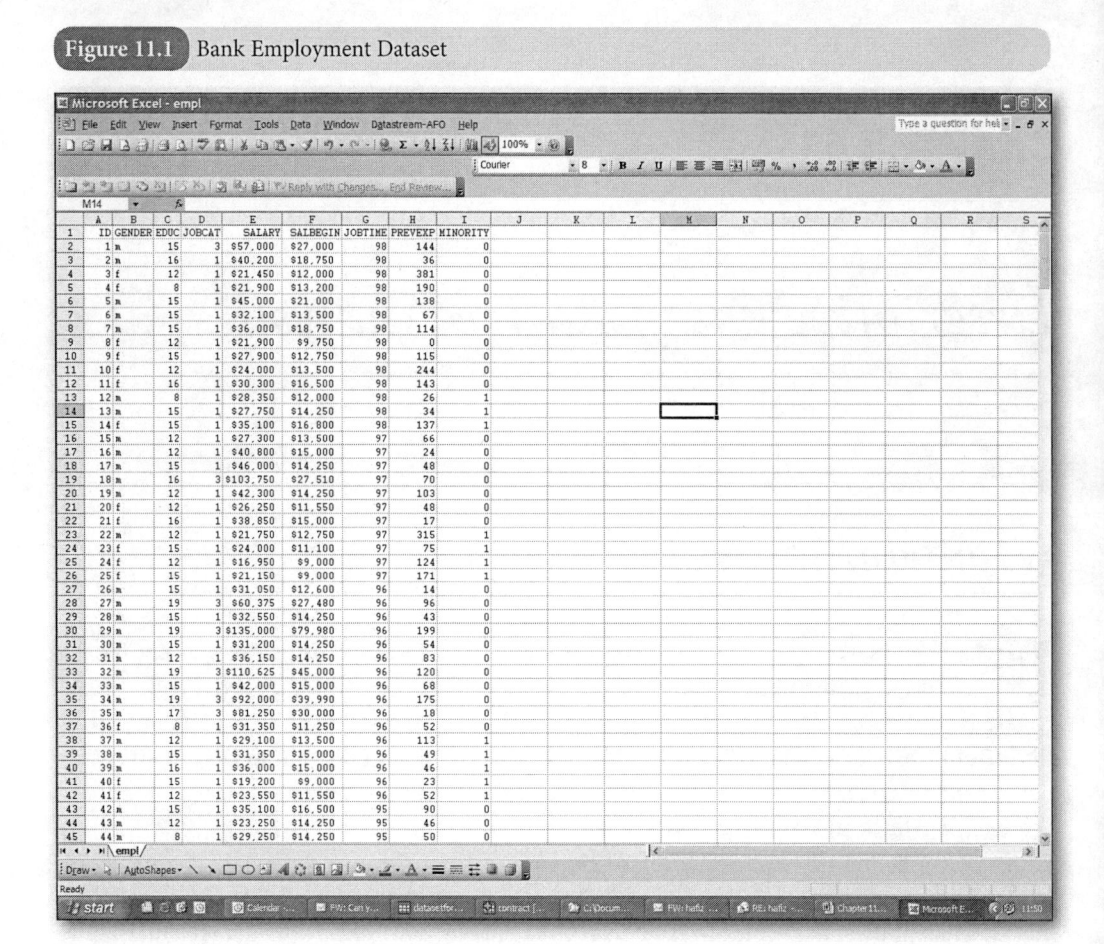

responses before entering into a computer data file. By definition, editing is the process of checking and adjusting the data for omissions, legibility and consistency.

Each of the columns shown in the worksheet in Figure 11.1 is called *variable* and each row is a *case*, which presents details pertaining in this situation to a particular employee. There are two types of variables. The first is continuous variables—entities that can be measured on a continuous scale, like months of previous experience (PREVEXP) and salary. The second is categorical variables, which can be further divided into nominal and ordinal variables. Nominal variables are variables in which class is arbitrarily assigned such as car colour or gender. Ordinal variables are those in which the level of the category, like social class, is defined, which implies higher or lower status. For the categorical variable job, category 1 is the code for a clerical worker, 2 is for security and 3 represents managerial level. Minority is an example of another categorical variable where 0 stands for 'not in a minority' and 1 is a 'minority group'.

11.2 DESCRIPTIVE STATISTICS

Descriptive statistics help us understand and summarise the data. The summary statistics can be represented either by tabular form or graphically. These useful tools are briefly discussed here.

Frequency and Frequency Distribution

Frequency is the count of each category in a certain variable. This count is often expressed as a percentage form or cumulative percentage form. For example, if we wish to know how many persons of each gender are employed by the bank, the following analysis can be used. In Excel, click on 'Insert' tab on the ribbon then 'Pivot Tables' (see Figure 11.2).

Move 'Gender' to column fields and also ID in to 'Drop Column Fields Here'. Click on top corner box and set to count. In gender variable 'm' stands for male and 'f' for female. This gives counts displayed in Table 11.1.

To get percentages, click on 'Count of Gender' and click on 'Options', then in the 'show values as' tab, click on 'No Calculation', then request row percentages of total. This gives the counts displayed in Table 11.2.

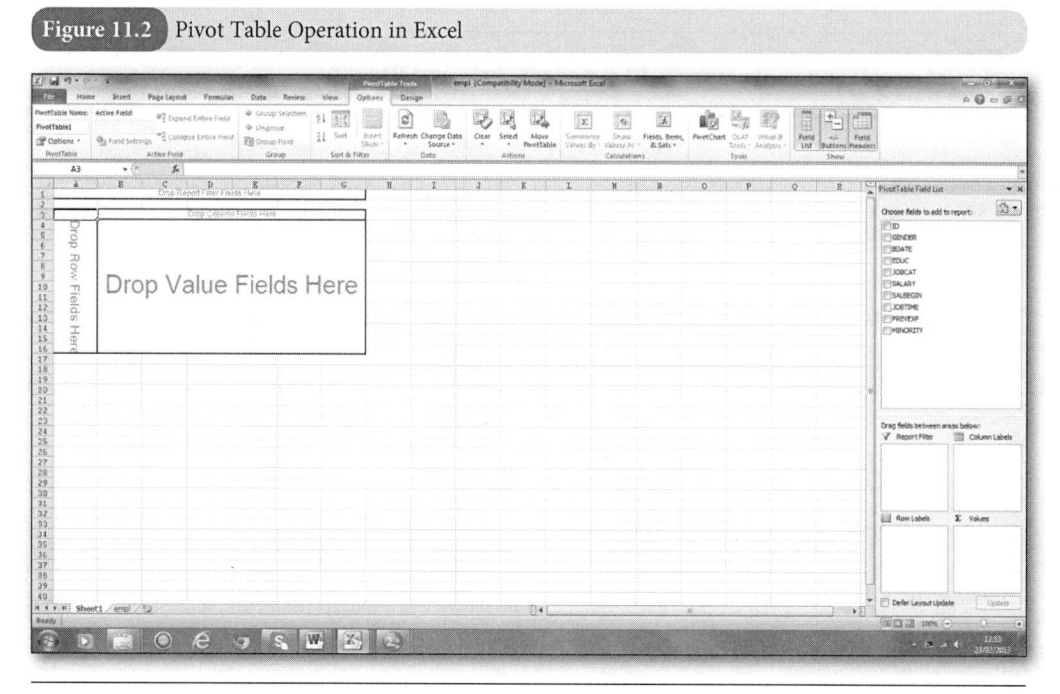

Figure 11.2 Pivot Table Operation in Excel

SOURCE: Authors' own.

Now drag Jobcat into Drop Column Fields here to get a two-way table; this gives Table 11.3. The clustered bar chart allows one to see the distribution of job type by gender.

This can be represented by a clustered bar chart showing row percentages as displayed in Figure 11.3.

What does this tell you about male and female employment in the bank?

Visual displays of data are very important to facilitate understanding and indeed a famous engineering statistician John Tukey (1986) stated, "display is an obligation".

Table 11.1 Frequency of Males and Females in the Data Set

Count of ID	Gender		Grand Total
	f	m	
Total	216	258	474

SOURCE: Authors' own.

Table 11.2 Percentages of Males and Females in the Data Set

Count of ID	Gender		Grand Total
	f	m	
Total	45.57%	54.43%	100.00%

SOURCE: Authors' own.

Table 11.3 Cross Tabulation of Gender by Job Category

Count of ID	Gender		Grand Total
JOBCAT	f	m	
1	206	157	363
2		27	27
3	10	74	84
Grand total	216	258	474

SOURCE: Authors' own.
NOTE: 1 = Clerical, 2 = Custodial and 3 = Managerial.

Figure 11.3 Employment Category by Gender

(Bar chart showing percentages for female (f) and male (m) across Clerical, Security, and Managerial categories. Clerical: f ≈ 57%, m ≈ 43%. Security: m ≈ 100%. Managerial: f ≈ 12%, m ≈ 88%.)

SOURCE: Authors' own.

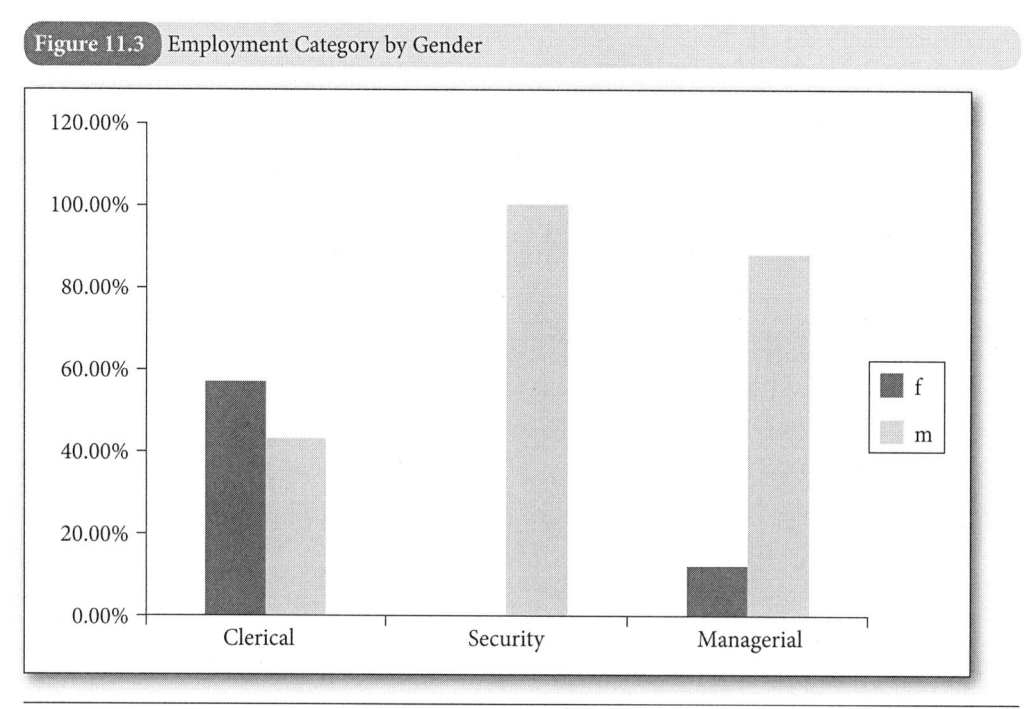

Measures of Location and Spread

To summarise continuous data, one quotes the most representative number usually as the average or mean ($\bar{X}$) and the median and measures of dispersion around the mean.

Consider the following set of numbers:

7   5   2   5   8   6   4   3   5

The *mean* is the sum of the numbers divided by the number of numbers, that is, 45/9 = 5.

If one arranges the numbers in order of size, the *median* is the middle number.

2   3   4   5   5   5   6   7   8

In this case the mean is also 5.

Linked to the median are *quartiles*; these split the number string into four groups. In the above example the lowest quartile is at 3.5, the middle quartile is the median and the upper quartile is at 6.5.

A measure of spread is the *range*, which is the difference between the maximum and minimum number, that is, 8 − 2 = 6.

A measure of spread which is not distorted by extremes is the *interquartile range*, which is the difference between the upper quartile and the lowest quartile, that is, 6.5 – 3.5 = 3.

A widely used measure is the *standard deviation(s)*, which is given by the following formula:

$$s = \frac{\sum_{i=1}^{n}(X_1 - \bar{X})^2}{n-1}$$

where s stands for the sample standard deviation, $\bar{X}$ is the mean of the observations, X_i is a particular observation I and n is the sample size.

The computation for the following data is displayed in Table 11.4.

7 5 2 5 8 6 4 3 5

The square of the standard deviation is known as *variance*.

The bigger the standard deviation, the more variable or dispersed the data is and it is a basic tool used for measuring the spread within individual variables. To conduct this analysis in Excel, go to DATA and then DATA ANALYSIS (this may need to be added in by choosing ADD INS then analysis tool pack). Then, go to descriptive statistics and enter the range of the data. A descriptive summary of the current salary (salnow) can be produced, as displayed in Table 11.5.

Table 11.4 Computation of the Standard Deviation

X_i	$(X_i - \bar{X})$	$(X_i - \bar{X})^2$
7	2	4
5	0	0
2	–3	9
5	0	0
8	3	9
6	1	1
4	–1	1
3	–2	4
5	0	0
$\bar{X} = 5$		Sum = 28
		$s^2 = 3.5$,
		$s = 1.87$

SOURCE: Authors' own.

Table 11.5 Descriptive Summary of the Variable Current Salary

Current Salary	
Mean	34,419.57
Standard error	784.31
Median	28,875
Mode	30,750
Standard deviation	17,075.66
Sample variance	291,578,214.45
Kurtosis	5.38
Skewness	2.12
Range	119,250
Minimum	15,750
Maximum	135,000
Sum	16,314,875
Count	474
Confidence level (95.0%)	1,541.17

SOURCE: Authors' own.

Let us only concern ourselves with the mean, median and standard deviation.

A graphical summary of this data is the *histogram*, which is shown in Figure 11.4.

To create a histogram, first define a 'bin range' in an empty column. This is the range of salaries (in the dataset we have put this in column k of the spreadsheet employe.xls).

Then in Excel go to Data Analysis and request Histogram—enter the appropriate fields, that is, f1:f475 and k1:k12—this gives bin and frequency columns—from which you can create a column chart. It is conventional not to have spaces between the columns—this is achieved by clicking on the column and then 'options' and reducing the space between columns to zero.

This is a skewed distribution. Often one may wish to make the distribution more symmetrical or normally shaped by using a transformation. This can be done by taking natural logarithms of the current salary. This is done by using the formula $= \ln(f2)$—see column L in the Excel spreadsheet. Salary now appears on a scale from 9.6 to 11.9. Log transformations are frequently applied to data, which ranges over a wide scale such as salaries or sales volumes or economic data such as gross domestic product.

If one wishes to compare male and female salaries, the above statistics can be used. To do this, first, sort the salaries in order of gender by going to 'data' and sort by gender. Then use 'Descriptive Statistics' twice to obtain Table 11.6.

Figure 11.4 Creation of a Histogram for Salary Data

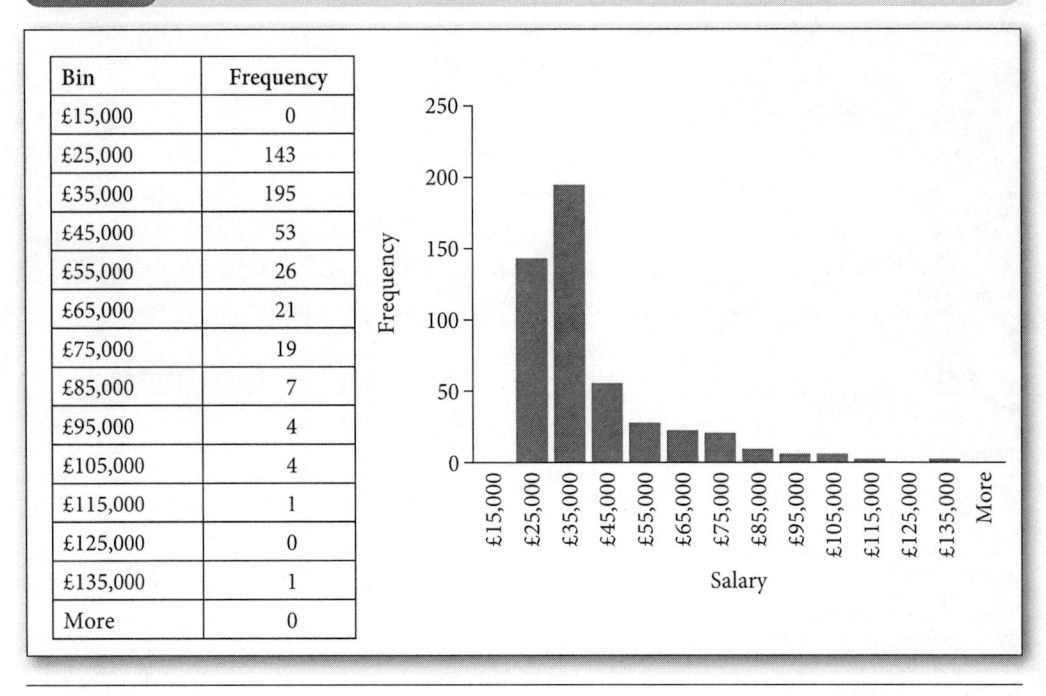

Bin	Frequency
£15,000	0
£25,000	143
£35,000	195
£45,000	53
£55,000	26
£65,000	21
£75,000	19
£85,000	7
£95,000	4
£105,000	4
£115,000	1
£125,000	0
£135,000	1
More	0

SOURCE: Authors' own.

To graphically compare groups *boxplots* are helpful. Boxplots (see Figure 11.5) are charts showing where 50 per cent of the data lie, that is, between the lower quartile and the upper quartile—this is the box. The whiskers run from the box to the lowest and highest observed data points, which are expected. Values beyond the ends of these whiskers are identified as outliers. These outlying points should be examined to ascertain their validity as part of the dataset. Boxplots are part of a set of techniques called exploratory data analysis or EDA— some of these techniques are outlined in Pelosi and Sandifer (2002). These techniques are not easily available in Excel and have been produced in SPSS (an Excel worksheet can easily be imported into SPSS).

Thus, males are paid more and their salaries are more variable. Pivot tables can also be used to get means, medians and standard deviations. There are a number of people receiving salaries greater than expected from the mass of the data. (In Figure 11.5 the circles indicate possible outliers and the stars indicate probable outliers.)

Returning to tables in Excel if one puts job category as columns, gender as rows and salary in the table, one obtains Table 11.7.

To change the statistic that is displayed in the table, click on the first cell (topmost corner) that appears in the pivot table. This initially appears as Sum of Salary and has to be changed to average.

Table 11.6 Comparison of Female and Male Salaries

	Female	Male
Mean	26,053	41,441.78
Standard error	516	1,213.97
Median	24,300	32,850
Mode	24,450	30,750
Standard deviation	7,569	19,499
Sample variance	57,292,062	380,219,336.30
Kurtosis	5	2.78
Skewness	2	1.64
Range	42,375	115,350
Minimum	15,750	19,650
Maximum	58,125	135,000
Sum	3,451,292.00	10,691,980
Count	194.00	258
Confidence level (95.0%)	1,151.57	2,390.59

SOURCE: Authors' own.

Table 11.7 Comparison of Female and Male Salaries

	Jobcat			
Average of Salary				
Gender	Clerical	Custodial	Manager	Average
Female	£25,003.69	—	£47,213.50	£26,031.92
Male	£31,558.15	£30,938.89	£66,243.24	£41,441.78
Grand total	£27,838.54	£30,938.89	£63,977.80	£34,419.57

SOURCE: Authors' own.

Scatter Plots

Plots are useful in examining how one continuous variable may depend on another. For example, to examine if salary is related to years of education, a simple scatter plot is helpful (see Figure 11.6).

Figure 11.5 Boxplots of the Distribution of Female and Male Salaries

SOURCE: Authors' own.

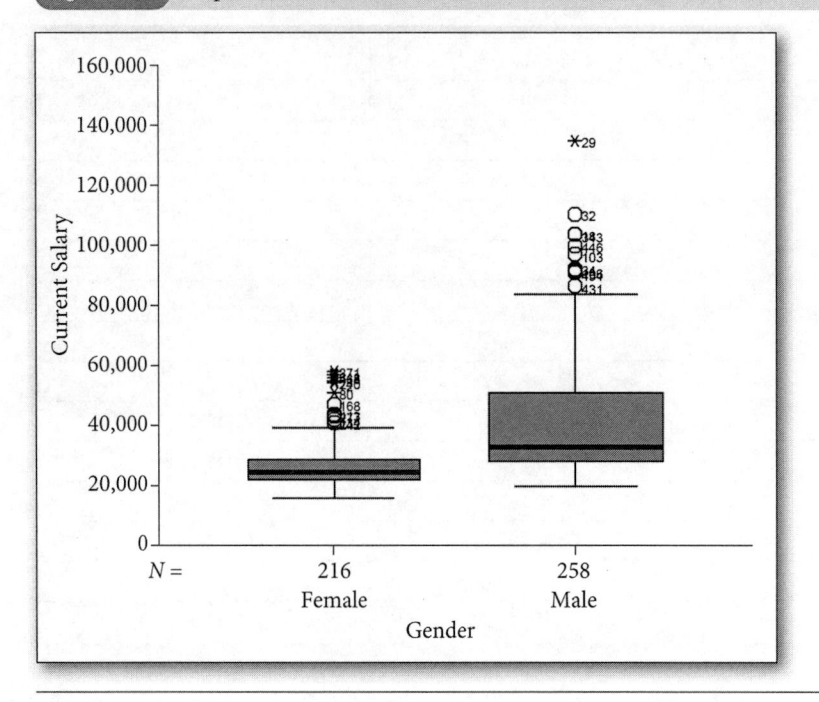

Figure 11.6 Salary Plotted against Age of Subject

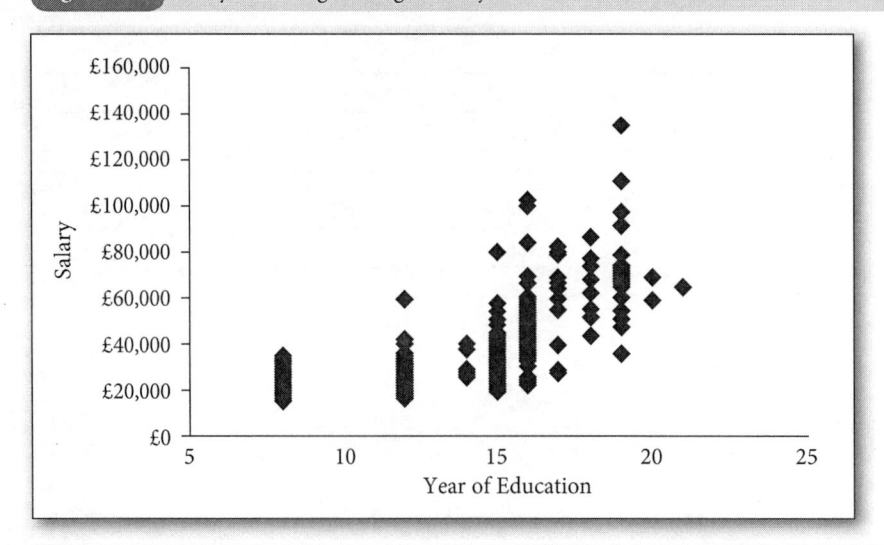

SOURCE: Authors' own.

More on Tables

Tables are often ignored but they are extremely useful data summary and analysis tools.

For example, British Oil's coal sales, in thousand tonnes, for its major markets in 1991 were:

USA 13,256; Germany 2,272; Rest of Europe 2,501; South Africa 3,514; Australia 4,505.

Using the above data, answer the following:

1. Which region was the largest coal market?
2. Which region was the smallest market for coal?
3. What is the difference in coal sales between Australia and the Rest of Europe?
4. What is the total coal sale?

These questions are easily answered if the data is displayed in a table such as Table 11.8.

Consider Table 11.9, which shows annual income by gender and age group.

What do you conclude from this table and what other information is required?

Another table (Table 11.10) could be the log of experimental results of yields produced at different temperatures.

Coding makes tables easier to interpret.

In this case use the transformation of divide the number in each cell of Table 11.10 by 10 and then subtracting 7 from the new number to get Table 11.11.

One can take a pencil and join numbers of the same level—this produces contours and one can determine that the best operating range is temperature levels 4–6 and pressure levels 4–6.

Table 11.8 British Oil Coal Sales by Market for 1991

Market	Coal Sales
USA	13,256
Australia	4,505
South Africa	3,514
Rest of Europe	2,501
Germany	2,272
Total	26,048

SOURCE: *Annual Company Report British Oil* (1992).
NOTE: Sales are shown in thousand tonnes.

Table 11.9 Distribution of Annual Income (in thousands), Selected by Gender and Age

Gender	Parameter	Age				
		16–20	21–30	31–40	41–50	51–60
Male	Number	20	60	40	30	20
	Mean	7	15	24	27	33
	SD	1.3	3	5	6	8
Female	Number	15	50	40	22	18
	Mean	6	12	18	20	10
	SD	1.4	3	5	6	6

SOURCE: Authors' own.

Table 11.10 Yields of a Process as Temperature and Pressure Are Varied

Pressure Level	Temperature							
	1	2	3	4	5	6	7	8
9	38	42	50	50	50	48	45	40
8	50	53	56	58	63	58	50	45
7	70	74	77	80	85	80	70	60
6	79	80	83	85	95	90	70	67
5	79	85	90	95	99	90	75	72
4	75	86	90	94	96	88	75	70
3	67	80	85	92	94	90	75	67
2	62	75	80	87	89	85	70	66
1	55	68	74	84	86	80	70	65

SOURCE: Authors' own.

Tabular Analysis

To illustrate tabular analysis further consider infant mortality by year and social class. The data is tabulated in Table 11.12.

Median Polishing

The effects of the passage of time and social class can be investigated by using a technique called median polishing. To apply this first subtract row medians and then column medians from the cell values in the respective row and column to get Table 11.13.

Table 11.11 Coded Yields of a Process as Temperature and Pressure Are Varied

Pressure Level	Temperature							
	1	2	3	4	5	6	7	8
9	-4	-3	-2	-2	-2	-2	-2	-3
8	-2	-2	-1	-1	-1	-1	-2	-2
7	0	0	1	1	2	1	0	-1
6	1	1	1	2	3	2	0	0
5	1	2	2	3	3	3	1	0
4	1	2	2	2	3	2	1	0
3	0	1	2	2	2	2	1	0
2	-1	1	1	2	2	2	0	0
1	-1	0	0	1	2	1	0	0

SOURCE: Authors' own.

Table 11.12 Infant Mortality Tabulated by Year and Social Class

Year	Social Class				
	1	2	3	4	5
1911	55.00	98.00	125.00	125.00	152.00
1921	38.00	55.00	78.00	89.00	97.00
1930–1932	32.00	45.00	57.00	66.00	77.00
1939	26.00	34.00	44.00	51.00	60.00
1945–1950	18.00	21.00	28.00	34.00	39.00

SOURCE: Authors' own.

The boxed number (57) is the column median of the row medians and is referred to as the 'common effect'. To predict a value in a particular cell ij, the following expression can be used:

$$(\text{data})_{ij} = \text{common} + \text{row effect}_i + \text{column effect}_j + \varepsilon_{ij}$$

For example, to predict infant mortality for the year 1921 and social class 2 would be found from 57 + 21 − 12 = 66. Thus an error or residual is obtained as the actual value: (55) − the prediction (66) = −11.

The median infant mortality rate is 57 and by converting the original data to this base it is easier to see the social class and year effects. Residuals are generally smaller than the effects. Converting the information to graphical presentations is very useful and is

Table 11.13 Table 11.12 with Column and Row Medians Subtracted

Row Medians					
125	−70	−27	0	0	27
78	−40	−23	0	11	19
57	−25	−12	0	9	20
44	−18	−10	0	7	16
28	−10	−7	0	6	11
57	−25	−12	0	7	19
Column Medians					
68	−45	−15	0	−7	8
21	−15	−11	0	4	0
0	0	0	0	2	1
−13	7	2	0	0	−3
−29	15	5	0	−1	−8

SOURCE: Authors' own.

displayed in Figure 11.7. Thus, infant mortality has decreased over time and the social class differentials have narrowed.

Questionnaires and Tables

Consider a situation in which 100 males and 50 females were asked to rate their views on nuclear energy on a five-point scale from very bad to very good. The distribution might be as shown in Table 11.14.

Computing row percentages makes the analysis easier (see Table 11.15).

If one thinks of the scale of very bad to very good being coded as the following weights:

$$-2 \quad -1 \quad 0 \quad 1 \quad 2$$

If for both males and females the percentage responses are multiplied by these weights and summed, one gets the following nominal scores:

Male Nominal Score: $-2 * 0 + (-1) * 20 + 0 + 40 + 2 * 10 = 40$
Female Nominal Score: $-2 * 40 + (-1) * 20 + 0 + 20 + 0 = -80$

Figure 11.7 Variation in Infant Mortality (INFMORT) with Year and Social Class

SOURCE: Authors' own.

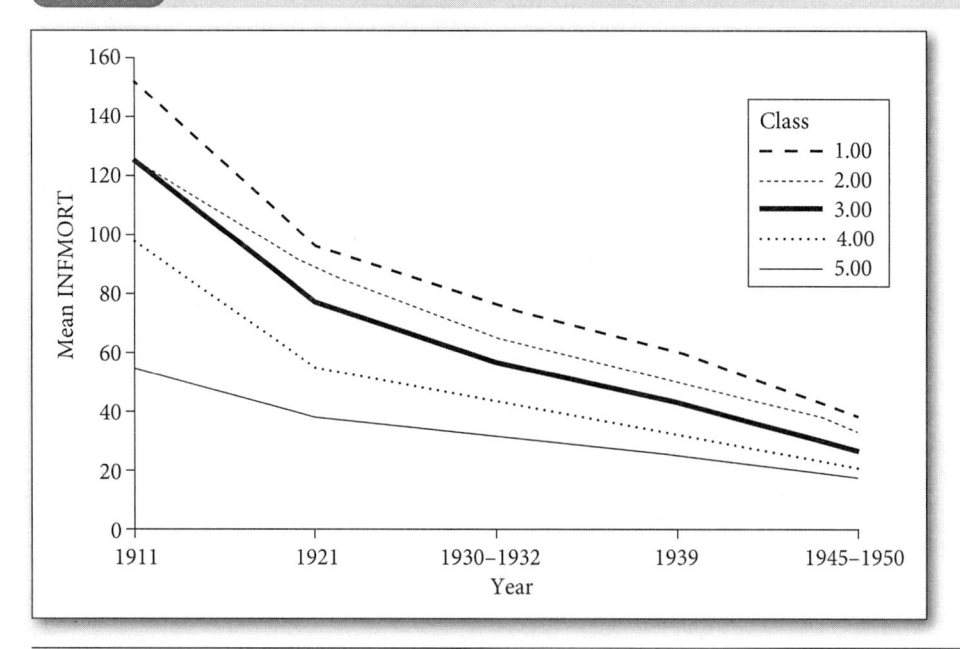

Table 11.14 Distribution of Question Responses

	Very Bad	Bad	Neutral	Good	Very Good
Males	—	20	30	40	10
Females	20	10	10	10	—

SOURCE: Authors' own.

Table 11.15 Row Percentage Distribution of Question Responses

	Very Bad	Bad	Neutral	Good	Very Good
Males	0	20	30	40	10
Females	40	20	20	20	0

SOURCE: Authors' own.

Thus, there is a clear difference between males and females. Consider the questions associated with factors which influence the use of the Internet. Subjects might have been asked to complete the following questions housed in Table 11.16.

The counts and distribution of those who answered this question are displayed in Table 11.17. The average response is also worked out and appears in the last column of the table.

From this a column chart of the averages can be obtained as displayed in Figure 11.8.

Thus, work, email and leisure are the main reasons for Internet use.

Table 11.16 Questions on Reasons for Internet Use

Item	View				
	1 Very Unimportant	2	3 Neutral	4	5 Very Important
Work					
Home					
Games					
Email					
Study					
Research					
Leisure					
Other					

SOURCE: Authors' own.

Table 11.17 Summarised Responses to Internet Use Questionnaire

Item	View					
	1	2	3	4	5	Average (Mean)
Work	0	2	4	17	12	4.11
Home	0	3	17	6	9	3.60
Games	7	14	5	6	13	3.09
Email	0	1	5	9	20	4.37
Study	5	5	16	5	4	2.94
Research	6	7	17	3	2	2.66
Leisure	0	5	6	11	13	3.91
Other	5	14	9	4	3	2.60

SOURCE: Authors' own.

Figure 11.8 Bar Chart of Reasons for Internet Use

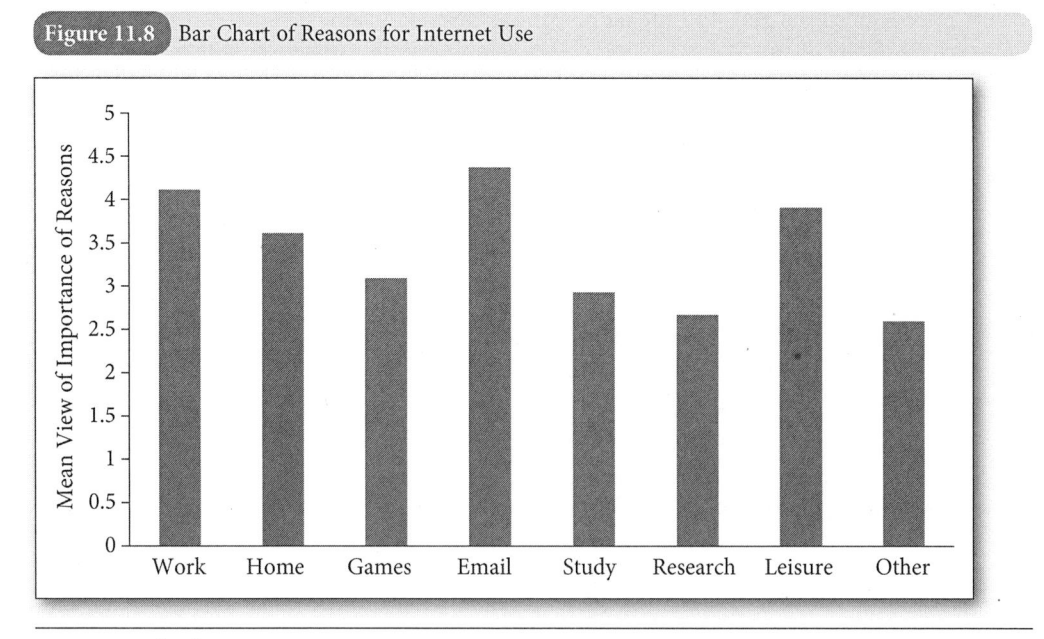

[Bar chart showing Mean View of Importance of Reasons on y-axis (0 to 5) for categories: Work (~4.1), Home (~3.6), Games (~3.1), Email (~4.4), Study (~2.9), Research (~2.7), Leisure (~3.9), Other (~2.6)]

SOURCE: Authors' own.

Guidelines for the Production of Tables

Guidelines for the production of tables can be listed as follows:

1. The table must have a clear purpose.
2. The table must have an explanatory title.
3. The table must clearly indicate the units of measurement of the data.
4. The table should clearly indicate the source of the data.
5. Lines should be drawn where appropriate to draw attention.
6. Row and column totals should be shown where meaningful.
7. Percentages and ratios should be computed if appropriate.
8. Don't present too many significant digits.
9. It is easier to look down a column of numbers than across a row.
10. Give attention to the spacing and layout of the table.

11.3 ARE THERE SIGNIFICANT DIFFERENCES?

Here, one asks, is there a significant difference between a group of sample measurements and a value? For example, in a statistical quality improvement situation, suppose in the manufacture of circuit boards the critical dimension is 80 microns. If a sample of five circuit boards is measured, the measures might be 81, 79, 76, 70 and 77 microns. The average of this sample is 76.6 microns. Is this a big enough difference from the target of 80 to say that the process is not meeting the target? To confirm this, another sample could be taken giving the results 78, 82, 88, 85 and 85 microns—giving an average of 83.6 microns; now what is to be concluded? Consider 100 samples of size five, which give the distribution of averages from the sample of size five (as shown in Figure 11.9).

The smooth curve as seen in Figure 11.9 is said to be the *Normal Distribution,* which theoretically is a distribution of where the averages would lie. The mean of all these samples is 75.1 microns and 95 per cent of the data lies between 74.7 and 75.5 microns. This is often referred to as the 95 per cent Confidence Interval of the mean. One is 95 per cent sure that the mean lies in this interval. The target of 80 does not lie in this interval, so it is concluded that the process is not conforming to target and it is significantly less than 80 microns. It appears that the second sample is significantly different from the rest of the data, as is apparently clear from the above histogram. Theoretically, for any sample the equation that gives the *95 per cent Confidence Interval* is

$$\bar{X} \pm 1.96 * \frac{s}{\sqrt{n}}$$

where *s* is the standard deviation, a measure of variability and *n* is the size of the sample. Thus, as the variability increases and *n* decreases, the confidence interval widens.

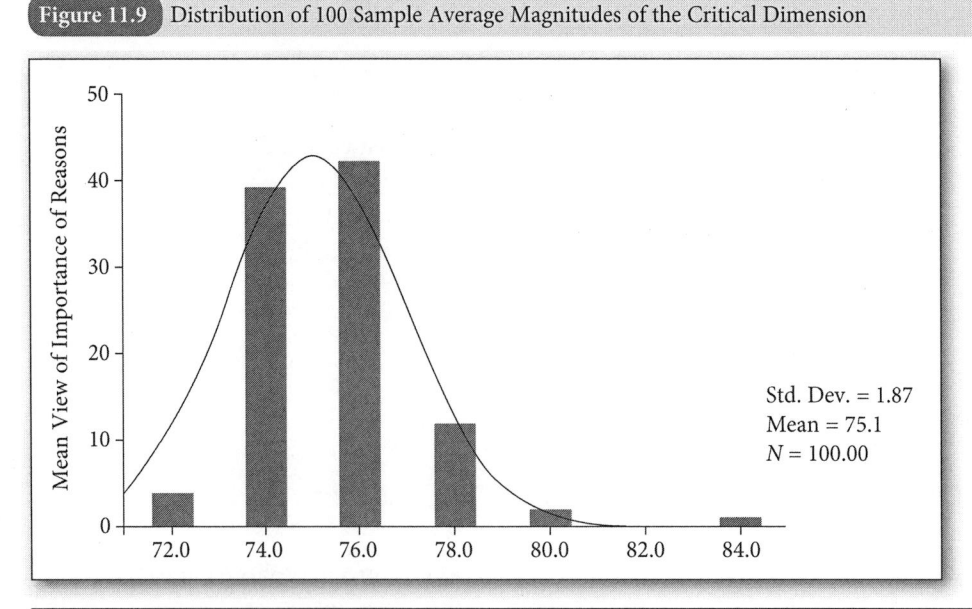

Figure 11.9 Distribution of 100 Sample Average Magnitudes of the Critical Dimension

SOURCE: Authors' own.

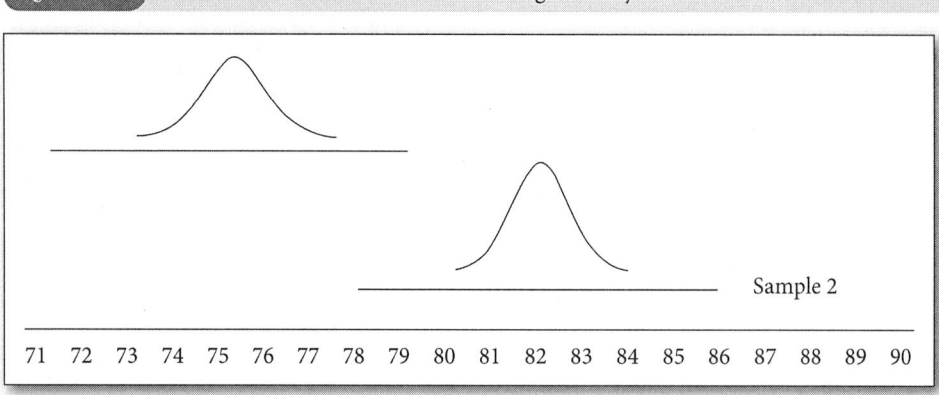

Figure 11.10 Two Normal Distributions which Are Significantly Different

```
                  ╱╲
                 ╱  ╲
    ────────────    ────────────
                              ╱╲
                             ╱  ╲
                     ────────    ──────────── Sample 2
    71  72  73  74  75  76  77  78  79  80  81  82  83  84  85  86  87  88  89  90
```

SOURCE: Authors' own.

For a sample of five readings, the first set has a confidence interval of 71.44–81.76 with a mean of 76.60 microns, while the second sample has a confidence interval of 78.9–88.30 with a mean of 83.60 microns. These confidence intervals overlap and so the two samples are said to *not* be significantly different. This is illustrated in Figure 11.10.

When a target is not included in the confidence interval then there is said to be a significant difference. Likewise if the two confidence intervals were not to overlap, then they would be significantly different.

If the confidence interval just bordered the target or another confidence interval then one would be 97.5 per cent sure that there is a difference. Why is that so?

Often these ideas are quoted as *P-values*, which is the likelihood of a difference occurring by chance—if this is less than 5 per cent (0.05) then this is taken as an acceptance of significance. (In SPSS, the *P*-value is called the *sig level*.)

Formally, one tests a null hypothesis (H_0) that there is no difference to an alternative hypothesis, (H_a) that there is a difference.

H_0: There is no significant difference.
versus
H_a: There is a difference (note there are three forms of this test: they are not equal, the mean is less than the target and that the mean is more than the target).

Comparing against a Target

Here, a one-sample *t*-test is used, which is available on the descriptive analysis section in Excel.

For example using the data in the file employe.xls one can ask the question: is the average years of education more than 10?

H_0: The mean is not different from 10.
versus
H_a: The mean is greater than 10.

To choose which hypothesis to accept, a one-sample t-test is used. Essentially, a confidence interval is constructed and if the test value is in that interval then H_o is accepted. Unfortunately, Excel does not readily execute the one-sample t-test. It does do two-sample t-tests, which are available in the data analysis menu. This can be used by putting a column of zero to compare with the variable of interest (years of education) at stating the level of difference from 0—which in this case is 10 (see Figure 11.11).

The output is given in Table 11.18.

Accept the alternative hypothesis as the p-value is less than 0.05; so 95 per cent sure that the mean is greater than 10.

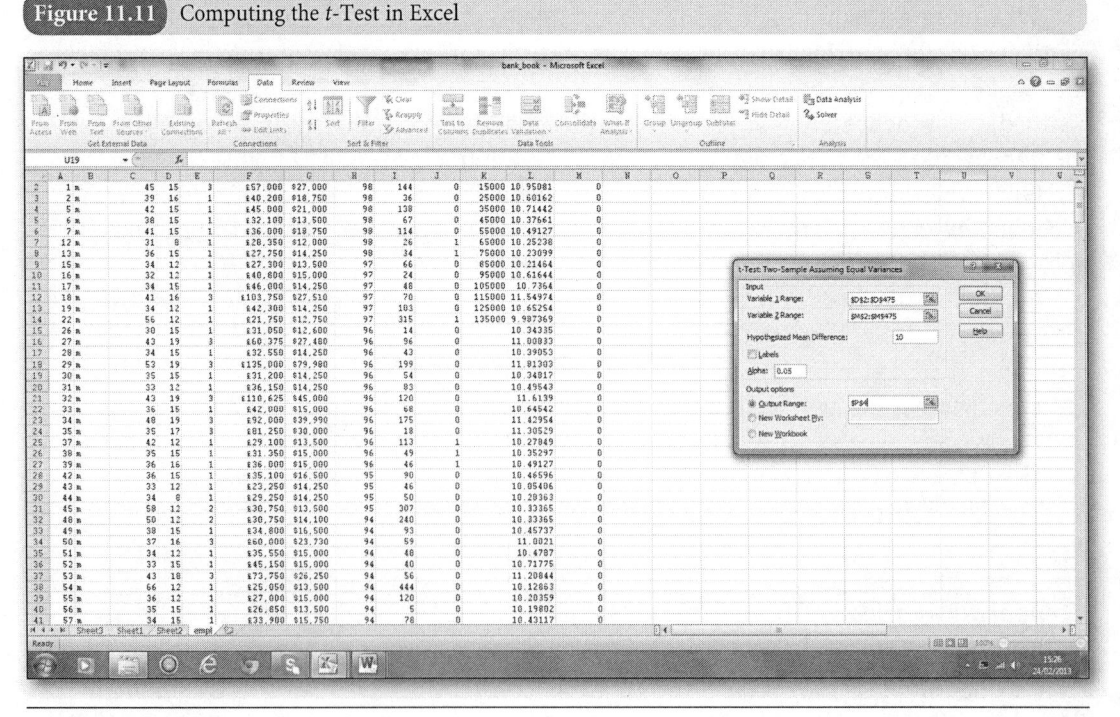

Figure 11.11 Computing the t-Test in Excel

SOURCE: Authors' own.

| Table 11.18 | *t*-Test on Mean of Years of Education |

	Years of Education
Mean	13.49
Variance	8.32
Observations	474
Pooled variance	4.16
Hypothesised mean difference	10
df	472
t-Stat	26.35
$P(T \leq t)$ one-tail	1.7E-115
t Critical one-tail	1.65
$P(T \leq t)$ two-tail	3.4E-115
t Critical two-tail	1.96

SOURCE: Authors' own.
NOTE: This is because $P(T \leq t)$ is less than 0.05 one accepts Ha and that the mean years of education is greater than 10.

11.4 COMPARING TWO GROUPS

Here, a two-sample *t*-test is used—there are two types: one when the variance (standard deviation squared) between the groups is broadly similar and the other when the variances are not similar. A rule for deciding this is that if the ratio of the standard deviation squared to the other standard deviation squared is greater than 2, then use the unequal variance version. In practice, it does not make much difference which one is used.

For example, consider the following: Are men paid more than females? For these *t*-tests, one requires data that is roughly symmetrical. So in this case it is best to use natural logarithms of the salary data.

Formally test,

H_0: the salary is not affected by gender.
versus
H_a: males are paid more than females.

To use the *t*-test, first sort the data set by gender and compute the standard deviations of log salary for each group and square these deviations. This gives the ratio $0.159/0.065 = 2.46$, so use unequal variance version of the *t*-test.

The output then is as displayed in Table 11.19.

Table 11.19 t-Test: Two-Sample Assuming Unequal Variances

	Females	Males
Mean	10.13	10.54
Variance	0.065	0.159
Observations	216	258
Hypothesised mean difference	0	
df	442	
t Stat	−13.63	
$P(T <= t)$ one-tail	7.38E-36	
t Critical one-tail	1.65	
$P(T <= t)$ two-tail	1.48E-35	
t Critical two-tail	1.97	

SOURCE: Authors' own.
NOTE: As p is less than 0.05 reject the null hypothesis and conclude that there is a difference.

Table 11.20 Performance Scores before and after Training

Individual	1	2	3	4	5	6	7	8	9	10	11	12
Before	55	61	47	65	58	57	55	50	65	61	60	55
After	60	68	53	60	70	63	53	58	68	64	60	69

SOURCE: Authors' own.

Paired Comparisons

This is used to compare two groups that only differ by one intervention.

For example, the performances of twelve individuals before and after a training intervention are displayed in Table 11.20.

H_0: Training makes no difference to performance.
versus
H_a: Training improves performance.

Using the paired samples *t*-test in the data analysis menu gives Table 11.21.

Thus, as p is less than 0.05 one can conclude that training has improved performance.

Table 11.21 t-Test: Paired Two Sample for Means

	Before	After
Mean	57.42	62.17
Variance	29.90	34.52
Observations	12	12
Pearson correlation	0.54	
Hypothesised mean difference	0	
Df	11	
t Stat	−3.02	
P(T <= t) one-tail	0.01	
t Critical one-tail	1.80	
P(T <= t) two-tail	0.01	
t Critical two-tail	2.20	

SOURCE: Authors' own.

11.5 COMPARING MORE THAN TWO GROUPS

To compare more than two groups, One Way Analysis of Variance (ANOVA) is used. For example, consider the following: Does educational level vary significantly with job grade?

H_0: There is no difference between educational level and job grade.
versus
H_a: There is a difference with job grade and educational level.

First, sort the data by job grade and copy years of education associated with people in the different job grades into three consecutive columns in a new worksheet (see ANOVA in employe.xls). Then go to data analysis, then request ANOVA single factors and enter the data ranges (see Figure 11.12).

From this procedure we obtain Table 11.22.

This shows as the *p*-value is less than 0.05, the alternative hypothesis should be accepted and that there is a significant difference in the years of education of employees in different job grades. Managers have the most years of education, followed by clerical workers.

Figure 11.12 One-Way ANOVA in Excel

SOURCE: Authors' own.

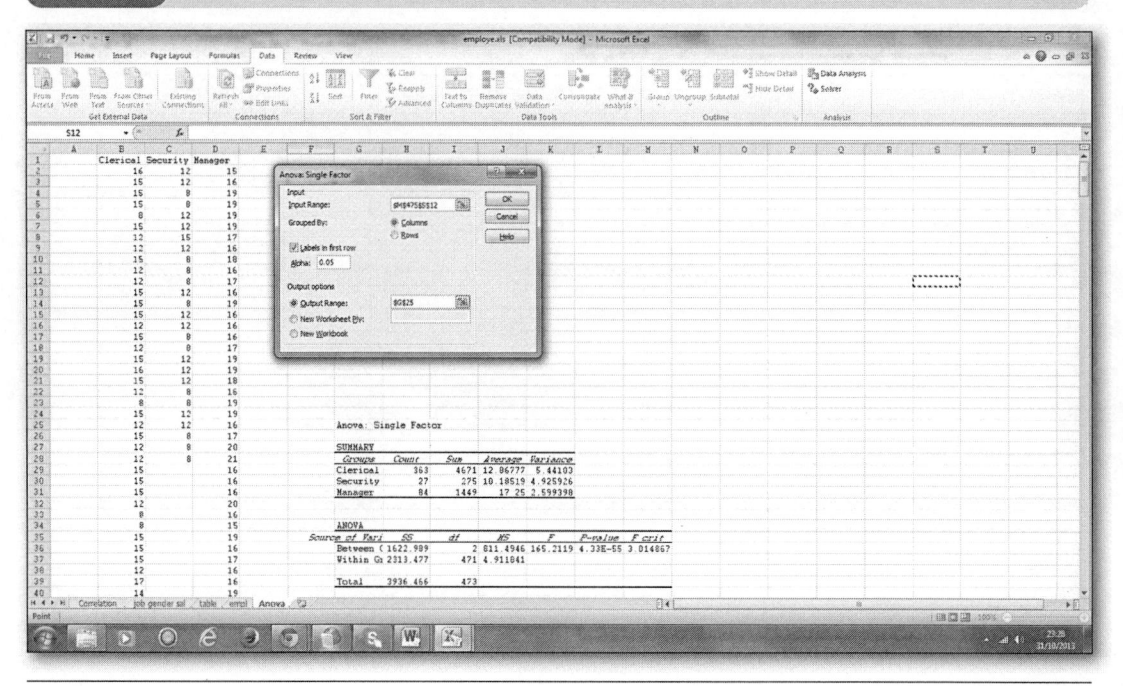

Table 11.22 ANOVA Single Factor

Summary

Groups	Count	Sum	Average	Variance
Clerical	363	4,671	12.87	5.44
Security	27	275	10.19	4.93
Manager	84	1,449	17.25	2.608

ANOVA

Source of Variation	SS	df	MS	F	P-value	F crit
Between groups	1,622.99	2	811.49	165.2119	4.33E-55	3.01
Within groups	2,313.48	471	4.91			
Total	3,936.47	473				

SOURCE: Authors' own.

11.6 THE ASSOCIATION BETWEEN CATEGORICAL VARIABLES

Contingency tables and Chi-square (χ^2) analysis are used to ascertain if categorical variables are associated. The procedure is illustrated for two categorical variables in this case: the variables are regional distribution and whether or not development grants are awarded. One might wish to investigate if grants are more likely to be awarded in one region rather than in another. One may have recorded the data shown in Table 11.23.

One wishes to test for association between columns (accept or reject) and the rows (region), that is,

H_0: there is no association.
versus
H_a: there is an association.

The expected distribution if the values in the table were distributed proportionately is computed from the row total multiplied by the column total all divided by the grand total.

Hence, for north region, the Expected Value is (362 × 216)/1,099 equals 71.15.

Putting the Expected Values in brackets, the above table becomes Table 11.24.

The Chi-squared statistic (χ^2) is now computed from:

χ^2 = the difference between observed and expected values in each cell squared and divided by the expected value and then all these are then summed.

If this is greater than the tabular value of the χ^2 distribution, then the alternative hypothesis is accepted.

Table 11.23 Regional Distribution of Grant Outcomes

Region	Application Successful	Application Rejected	Row Totals
North	63	299	362
East	55	207	262
West	44	208	252
South	54	169	223
Column total	216	883	1,099

SOURCE: Authors' own.

The tabular value is given by χ^2_α, $(r - 1)*(c - 1) = 7.815$; so reject the alternative hypothesis and conclude that there is no evidence in the awarding of development grants.

Note: For chi-square to be reliable, the number in the cell should be greater than 5 and there should not be a multiple of greater than 10 between cells.

In Excel, one can form the table using pivot tables but one has to use formulas to compute the expected values and the chi-square. In other computer packages such as SPSS, this is easily computed. An example is as follows:

Analyse > crosstabs > options > statistics

H_0: there is no association between variables.
  versus
H_a: there is an association.

For example, to test if there is an association between gender and minority classification chi-square analysis in the crosstab function of SPSS gives Tables 11.25 and 11.26.

Table 11.24 Observed and Expected Values of Regional Distribution of Grant Outcomes

Region	Application Successful	Application Rejected	Row Totals
North	63 (71.15)	299 (290.85)	362
East	55 (51.49)	207 (210.51)	262
West	44 (49.53)	208 (202.47)	252
South	54 (43.83)	169 (179.17)	223
Column total	216	883	1,099

SOURCE: Authors' own.

Table 11.25 Cross-Tabulation of Gender and Mortality with Expected Values

			Minority Classification		
			No	Yes	Total
Gender	Female	Count	176	40	216
		Expected count	168.6	47.4	216.0
	Male	Count	194	64	258
		Expected count	201.4	56.6	258.0
Total		Count	370	104	474
		Expected count	370.0	104.0	474.0

SOURCE: Authors' own.

Table 11.26 Chi-Square Analysis Outcomes

	Value	df	Asymp. Sig. (2-sided)	Exact Sig. (2-sided)	Exact Sign (1-sided)
Pearson chi-square	2.714[b]	1	.099		
Continuity correction[a]	2.359	1	.125		
Likelihood ratio	2.738	1	.098		
Fisher's exact test				.119	.062
No. of valid cases	474				

SOURCE: Authors' own.
NOTES: The number in each cell needs to be greater than 5.
    [a] Computed only for a 2×2 table.
    [b] Cells (.0%) have expected count less than 5. The minimum expected count is 47.39.

11.7 SUMMARY OF TEST PROCEDURES

The tests outlined in this chapter are summarised in Table 11.27.

11.8 EXERCISES

Using the data in the file employe.xls
Load file from http://www.sagepub.in/books/Book242499/samples.

1. Obtain a scatter graph of how salary depends on previous experience.
2. Obtain a clustered bar chart of percentage in job category by minority status.
3. Compare salaries for those in an ethnic minority with those who are not. Repeat this for each job category.
4. In Excel use the formulas = average (range), = median (range) and – st.dev (range) to find the median, average and standard deviation for salary and years of education. To find the average of salary go to a cell on the worksheet and enter = average (f2:f475).
5. Use pivot tables to compute the mean years of education for job category by ethnic minority.
6. In Table 11.28 the incidence of crime in a country is displayed.
   Calculate the row per cents and interpret the table.
   Is there any association between the crime and the region where the crime was committed?

Table 11.27 Summary of Test Procedures

Requirement	Example of Situation	Test to be Used	Comment	Theoretical Test
Compare to a target.	Is the average age of employees more than 40 years?	Use a one sample t-test.	Think carefully about hypothesis to be tested. Continuous data only.	If the data set is more than 30 use a z-test.
Compare two groups with no control.	Is male absenteeism less than female absenteeism?	Use independent samples t-test.	Decide first whether variances (standard deviation squared) is similar between groups. Apply either variances assumed equal or variance assumed unequal. Continuous data only.	If the data set is more than 30 use a z-test.
Compare two groups with one controlled intervention.	Test scores before and after training.	Use paired t-test.	Always question the validity that there has been only one intervention. Continuous data only.	
Compare more than two groups.	Compare amount of waste between four manufacturing plants.	Use One Way Analysis of Variance.	Continuous data only.	
To ascertain if there is an association between two categorical variables.	Is there an association between gender and job grade?	Contingency tables and Chi-square test.	All for categorical data.	

SOURCE: Authors' own.

Table 11.28 Crime by Region

Region	Crime Incidence		
	Homicide	Rape	Arson
South	8	30	25
West	50	230	280
East	70	350	180
North	15	161	541

SOURCE: Authors' own.

7. Use the data Table 11.29 to investigate if there is any association between social class and diagnostic category of happiness of psychiatric patients.

Table 11.29 Social Class and Patient Happiness

Social Class	Diagnosis			
	Depressed	Sad	Happy	Euphoric
1	46	25	18	15
2	12	45	28	22
3	10	12	12	10
4	3	8	20	16
5	2	4	22	20

SOURCE: Authors' own.

8. The data in Table 11.30 is the number of franchisors in a region by age and size of an enterprise. Apply median polishing and draw conclusions.

Table 11.30 Number of Franchisors by Company Age and Size

Age	Size				
	Micro	Small	Medium	Large	Total
Newly emerged	29	23	35	6	93
Young	18	6	14	3	41
Older	21	12	33	9	75
Mature	4	5	12	39	60
Total	72	46	94	57	269

SOURCE: Authors' own.

9. Using the data in the file employe.xls analyse the following: is there evidence that the mean number of months in employment is not 80?
10. Using the data in the file employe.xls compare the educational levels of whether or not someone is in a minority.
11. Using the data in the file employe.xls how does the salary of male and female clerical workers compare.
12. The percentage market penetration of a product in eight regions before and after advertising is tabulated in Table 11.31.

Table 11.31 Market Penetration by Region before and after Advertising

Advertising	Region							
	1	2	3	4	5	6	7	8
Before	12	15	10	25	21	44	21	25
After	15	25	8	30	33	49	20	30

SOURCE: Authors' own.

Use a paired t-test to determine if penetration has improved significantly.

13. Conduct a Chi-square test for Table 11.32.

Table 11.32 Distribution of Employees by Gender and Job Grade

	Number in Job Grade		
	Shop Floor	Supervisor	Manager
Female	102	20	10
Male	200	60	25

SOURCE: Authors' own.

11.9 REFERENCES

Bryman and D. Crammer, *Quantitative Data Analysis with SPSS 12 and 13: A Guide for Social Scientists* (London: Routledge, 2004).

A. Field, *Discovering Statistics Using SPSS* (Third edition) (London: SAGE Publications, 2009).

M.K. Pelosi and T.M. Sandifer, *Doing Statistics for Business with Excel: Data, Inference and Decision Making* (Second edition) (New York: John Wiley and Sons, 2002).

J. Tukey, *The collected Works of John W. Tukey* (Belmont CA: Wadsworth and Brooks/Cole, 1986).

CHAPTER 12

Correlation and Regression

12.1 INTRODUCTION

In this chapter we shall consider measuring the association between variables and, if they are linked theoretically, how one can build a model of the relationship. We will begin with only two variables and later in the chapter more variables will be introduced.

12.2 CORRELATION

Pearson's Product Moment Correlation Coefficient (r) is a measure of the degree of *association* between variables. It takes a value between −1 and 1. A value of r close to 1 indicates strong positive association, for example, a person's height and weight; whereas a value of r close to −1 indicates a strong negative linear association, for example, amount of alcohol consumed and performance in a test. When $r = \pm$ this indicates that the two variables are perfectly correlated, that is, all the points are on a straight line. Some scatter plots with values of the Pearson product moment correlation coefficient are displayed in Figure 12.1.

Pearson's correlation coefficient is computed from

$$r_{xy} = \frac{\text{cov}(x,y)}{S_x S_y} = \frac{\sum (X_i - \bar{X})(Y_i - \bar{Y})/(n-1)}{S_x S_y}$$

This formula is available in most packages including SPSS. SPSS also gives a significance level; if this is less than 0.05, then the coefficient is judged significant. However,

Figure 12.1 Scatter Plots with Associated Correlation Coefficients

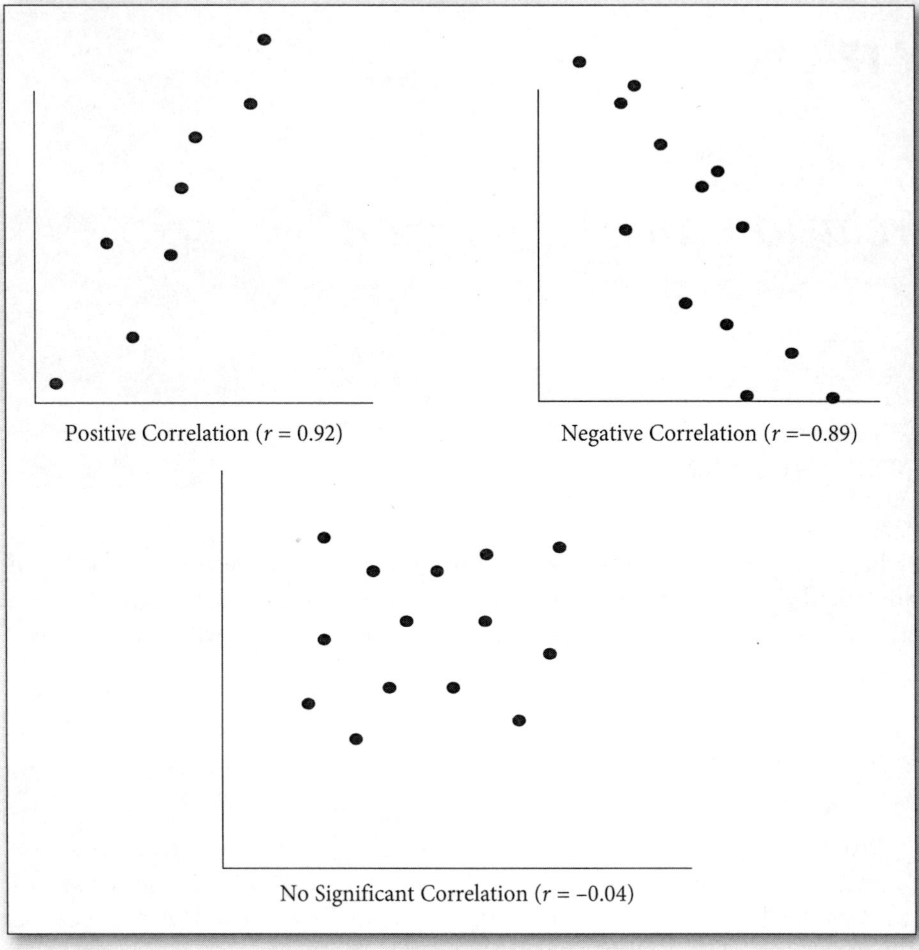

Positive Correlation ($r = 0.92$)

Negative Correlation ($r = -0.89$)

No Significant Correlation ($r = -0.04$)

SOURCE: Authors' own.

one should not consider the coefficient on its own; scatter plots should also be examined. Figure 12.2 illustrates the dangers of ignoring the plots.

As an example of calculating correlations using Excel, load the file *hdr2012.xls* (this file can be downloaded from http://www.sagepub.in/books/Book242499/samples) and go to *DATA* and *Data Analysis* and then *Correlation* and enter the variables ln(GDP/_PC) (natural logarithm of gross domestic product per capita in 2011), TFR 2011 (total fertility rate in 2011), ALR 2010 (adult literacy rate in 2010), FLE (female life expectancy at birth in 2012), IMR 2010 (infant mortality rate in 2010) and LE 2011 (life expectancy at birth in 2011).

Figure 12.2 Misleading Correlations

Outlier effect turns $r = 0.002$ to $r = 0.91$. The outlying point should perhaps be removed.

Non-linear relationship ($r = -0.02$). Pearson's correlation coefficient is purely a measure of linear association. Yet many variables have strong non-linear association, such as economies of scale.

SOURCE: Authors' own.

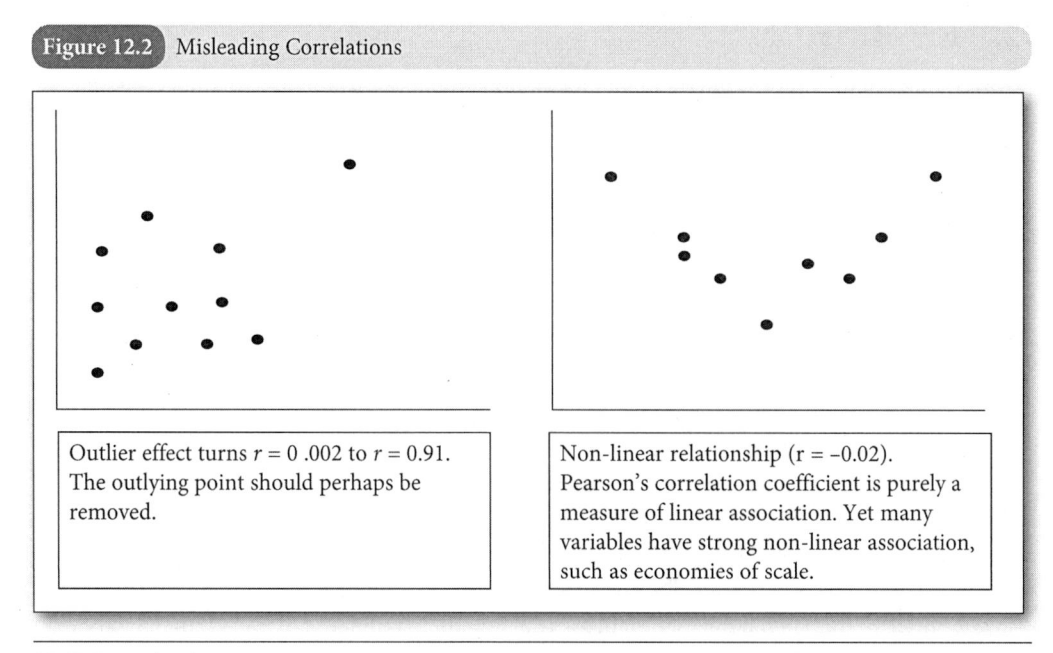

Table 12.1 Correlation Matrix

	ln(GDP-PC)	TFR 2012	ALR 2010	IMR 2010	LE 2011
ln(GDP-PC)	1				
TFR 2012	−0.758	1			
ALR 2010	0.756	−0.764	1		
IMR 2010	−0.808	0.838	−0.807	1	
LE 2011	0.790	−0.819	0.720	−0.928	1

SOURCE: Authors' own.

A Correlation Matrix for the selected four variables is computed using the correlation function in Excel, which is available in the Data Analysis tools in the DATA list.

The output is shown in Table 12.1. This data was extracted from the statistics section of the UN's Human Development Report—see http:// http://hdr.undp.org/en/.

A matrix scatter plot is also helpful, but unfortunately Excel software does not provide such a diagram. If you have access to SPSS, you could investigate this. Thus, a strong positive correlation is suggested between Adult literacy and log of GDP and contraceptive prevalence. Negative correlations exist between log of gdp per capita and the total fertility rate and the infant mortality rate. Note that the infant mortality rate and total fertility rates are positively correlated.

12.3 REGRESSION

Regression is concerned about finding a relationship between variables and forming a model. Hence, this is a major tool of statistical modelling. An equation may be developed with independent variables, which have stimuli or influence on a dependent variable or response. This method was first developed by Legendre in 1805.

Assumptions

1. The relationship is sensible.
2. The relationship is linear.
3. Errors are independent and normally distributed.

Linear Regression is often used to explain changes in some phenomenon as a result of influencing variables. This involves estimating the coefficients of the explanatory variable or independent variables or stimuli that go to predict the dependent variable or response. Consider the example of a restaurant owner who thinks that sales of pizzas in cities are dependent primarily on the size of the student population. To test this, he gathers the data, which is tabulated in Table 12.2.

To test this, the data is first plotted on a scatter diagram (see Figure 12.3) and a visual assessment is made to determine if a straight-line relationship is suitable. In addition, the

Table 12.2 Annual Pizza Sales and Number of Students

City	Student Population (1,000s)	Annual Sales (£1,000s)
1	2	58
2	6	105
3	8	88
4	8	118
5	12	117
6	16	137
7	20	157
8	20	169
9	22	149
10	26	202

SOURCE: Authors' own.

Figure 12.3 Scatter Plot of Sales against Students (in Thousands)

SOURCE: Authors' own.

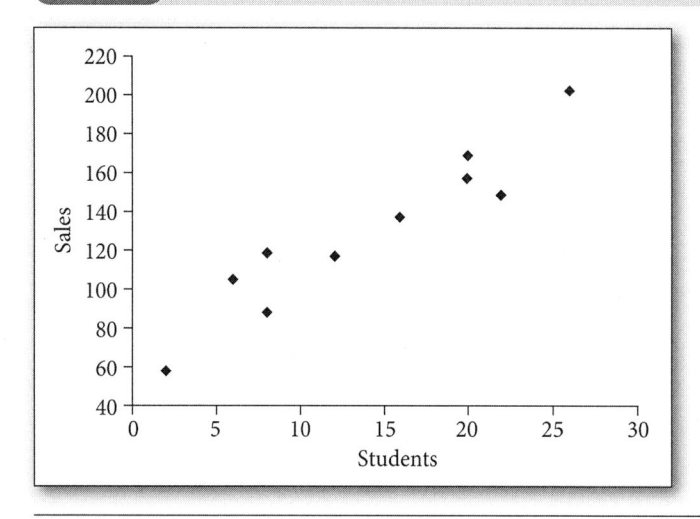

data is checked for values, which may not fit with the model—perhaps odd values might be data collection or inputting errors or they might be suggesting that straight-line models might not be appropriate.

It looks as though a straight-line positive relationship may well be appropriate. This is supported by the high positive correlation coefficient of 0.905, which is highly significant with a P value < 0.001. We now fit the 'best' straight line through the data and calculate the equation of the line. Regression is a way of doing this and estimates the coefficients or parameters of the model:

$$y_i = \beta_0 + \beta_1 x_i + \varepsilon_i$$

in such a way that the sum of squares of the errors or residuals is minimised, that is,

$$\text{Minimise} = \sum_{i=1}^{n} e_i^2 = \sum_{i=1}^{n} (y_i - \hat{y}_i)^2$$

The restaurant owner uses Excel to compute the equation by going to DATA, then Data Analysis, selecting Regression and entering 'sales' as the *dependent* and 'student population' as the *independent*. The owner then has the output shown in Table 12.3.

This gives the model: Pizza Sales = 60 + 5 * students, which mean that for every additional 1,000 students, annual pizza sales increase by £5,000. We now ask is this

Table 12.3 Simple Regression Analysis

Summary Output

Regression Statistics	
Multiple R	0.95012
R^2	0.90273
Adjusted R^2	0.89058
Standard error	13.82932
Observations	10

ANOVA

	SS	MS	df	F	Sig F
Regression	14,200	14,200	1	74.2484	0.0000
Residual	1,530	191.25	8		
Total	15,730		9		

	Standard Error	t-Stat	Coefficients	P-value	Lower 95%	Upper 95%
Intercept	9.2260	6.5033	60	0.0002	38.7247	81.2753
Students	0.5803	8.6167	5	0.0000	3.6619	6.3381

SOURCE: Authors' own.

model any good and is it reliable? The value R^2 is the coefficient of determination and is often used as a test of quality. This is calculated as follows:

$$R^2 = \frac{\text{Sum of Squares Explained by Regression}}{\text{Total Sum of Squares (before Regression)}}$$

$$= \frac{\sum (\hat{y}_i - \bar{y})^2}{\sum (y_i - \bar{y})^2}$$

In this case, R^2 is over 90 per cent. It suggests that 90 per cent of the variation in the sales of pizzas is explained by the model. Generally, one wishes R^2 to be as close to 100 per cent as possible. However, it is dangerous to rely solely on R^2. One must consider the assumptions behind the model and diagnose the model accordingly.

The assumptions to consider are:

1. The model makes theoretical sense.
2. The errors ε_i are independent of one another (uncorrelated) and normally distributed with mean zero.
3. The parameters are constant over time.

12.4 DIAGNOSTICS

To diagnose the model with respect to these assumptions, the following methods are available:

1. Is the relationship sensible?
2. Examine the signs of the coefficients—are they sensible?
3. Test the errors for a zero mean and for normality. To do this, one can look at a normal probability plot (normal P–P plot in Excel) of the residuals (see Figure 12.4). (The residuals are the observed values minus the fitted values.)

The residuals should always be plotted against the predicted values. One hopes that a random scatter will appear. The plot is displayed in Figure 12.5. They can also be plotted against the explanatory variable.

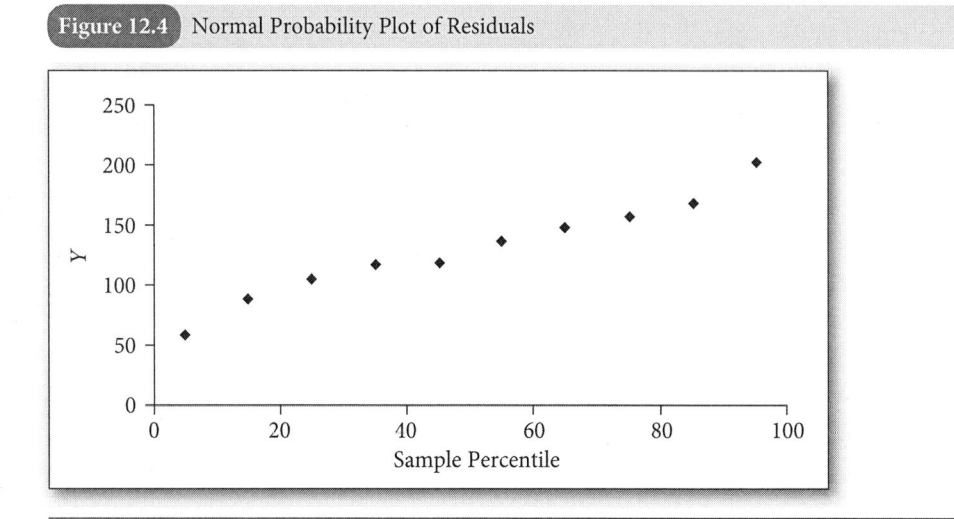

Figure 12.4 Normal Probability Plot of Residuals

SOURCE: Authors' own.

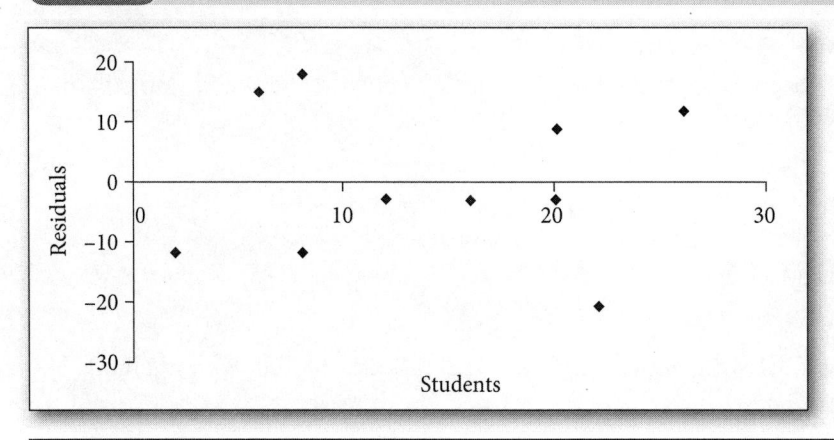

Figure 12.5 Residuals Plotted against Predicted Values

[Scatter plot of Residuals vs Students, y-axis from -30 to 20, x-axis from 0 to 30]

SOURCE: Authors' own.

This analysis allows correlation, non-constant variance (heteroscadisity) and influencing points to be detected. Fortunately, the residuals appear randomly scattered as was hoped for. Often one will work with the standardised residuals (SR). These are computed automatically from

$$SR = \frac{y_i - \hat{y}_i}{s}$$

where s is the standard deviation of the residuals. Absolute values of SR > 3 indicate a probable deviant requiring investigation. Absolute values of SR > 2 but < 3 indicate a possible deviant.

Are the Coefficients Significantly Different from Zero?

To assess this, estimated coefficients are compared to their standard errors and if significantly different from zero then the magnitude of coefficients should be more than twice their standard error.

More formally, the *t*-ratio is computed from

$$t\text{-ratio} = \left|\frac{\text{coefficient}}{\text{standard error of coefficient}}\right|$$

This is tested against the Student t distribution, that is, the t ratio is compared to

$$t_{\alpha/2, n-k}$$

where k = the number of parameters including β_0.
The hypothesis is as follows:

H_0: the coefficient is not significantly different from zero
versus
H_1: the coefficient is significantly different from zero

Reject the null hypothesis (H_0) if the T ratio is less than the table value.
The F test allows one to test if a linear form between y and x exists. The test is:

H_0: All the β's equal zero
versus
H_1: At least one of the β's do not equal zero

This is compared to the F-distribution.

$$F_{\alpha/2, p, n-k}$$

where p is the number of independent variables.
For both the above tests, p-values are computed automatically by most computer packages.

12.5 MULTIPLE REGRESSION

This is an extension of the previous section and the concept is to build models with *several* explanatory variables:

$$y = \beta_0 + \beta_1 x_1 + \beta_2 x_2 + \cdots + \beta_n x_n + \varepsilon$$

where y is the response and the independent variables are those that 'explain' the response. For example, demand for electricity might be explained by family size, amount of consumer durables, temperature, time of day, day of week and so forth. Determining, specifying and measuring these independent variables can be difficult.

Consider the case of a researcher investigating the hours of television watched per day by male pensioners. Data has been collected on the average number of hours watched, the pensioners' marital status (coded 0 if unmarried and 1 if married), their age and their years of education. The data collected is displayed in Table 12.4 and the variable names are shown in brackets in this table.

Mstatus is a special type of variable where a value is assigned to indicate one state or another, and is often called an *indicator* or a *dummy* variable. The dependent variable is graphed against age and years of education as shown in Figure 12.6.

Table 12.4 TV Watching by Pensioners

Hours Watched	Marital Status (Mstatus)	Age	Years of Education (Yrs_Ed)
0.5	1	73	14
0.5	1	66	16
0.7	0	65	15
0.8	0	65	16
0.8	1	68	9
0.9	1	69	10
1.1	1	82	12
1.6	1	83	12
1.6	1	81	12
2	0	72	10
2.5	1	69	8
2.8	0	71	16
2.8	0	71	12
3	0	80	9
3	0	73	6
3	0	75	6
3.2	0	76	10
3.2	0	78	6
3.3	1	79	6
3.3	0	79	4
3.4	1	78	6
3.5	0	76	9
3.6	0	65	12
3.7	0	72	12

SOURCE: Authors' own.

Figure 12.6 Hours of TV Watched Plotted against the Dependent Variable

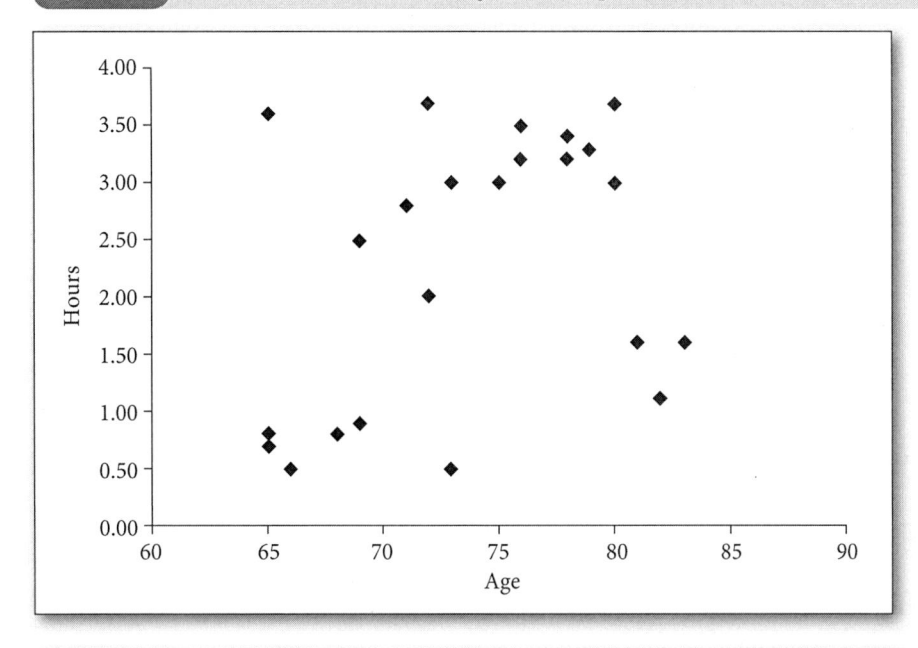

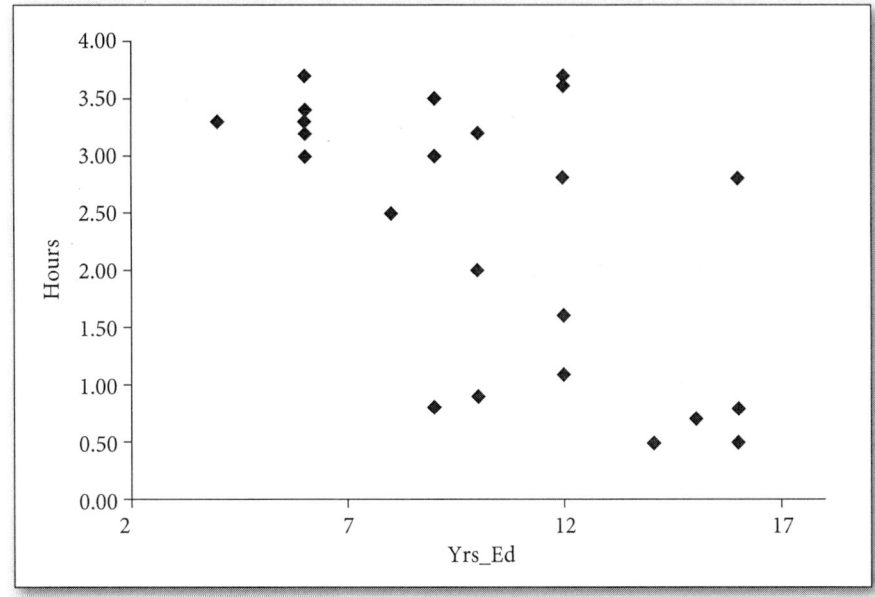

SOURCE: Authors' own.

This shows that there is a weak positive link between 'hours watched' and 'age'. There is a cluster of over 80-year olds who buck the trend—perhaps, these people are very ill. The variable 'hours watched' appears to drop with 'years of education'. Computing the average hours of TV watched per day of those unmarried and those who are married gives respective values of hours watched per day as 2.76 and 1.62 hours, respectively. The difference between the two groups is significant at the 5 per cent level as determined by an independent samples t-test. These relationships are confirmed when correlations are computed (see Table 12.5).

The view obtained from the graphs is supported. Next we wish to model this relationship by conducting regression modelling in Excel using regression in the data analysis tools (see Figure 12.7).

The coefficient of determination is 57 per cent. This is denoted by the Adjusted R^2 value in Table 12.6. Adjusted R^2 is used if there are more than one independent variables in the model as is the case with multiple regression. Clearly, there is variability still to be explained. What other variables would you wish to use?

This gives the output displayed in Table 12.6.

Thus, there is evidence of a significant relationship as both marital status and years of education have p-values lower than 0.05. However, the p-value for age is higher than 0.05 and this suggests that age is not a significant factor in explaining hours of TV watched. Marital status has the biggest effect, followed by years of education and, finally, age has the least effect. The normal probability of the residuals, displayed in Figure 12.8, suggests that the model is satisfactory in meeting assumptions relating to normality of the errors.

A key concept in modelling is the principle of parsimony, which is to try and explain the response as simply and with as few terms as possible. Therefore, age should be removed from the modelling process as it is not a significant factor. A simpler model is now produced and the output is presented in Table 12.7.

This means that those who are married and have high education tend to watch less TV. The diagnostic plots are displayed in Figure 12.9 and indicate a reasonably satisfactory fit.

Table 12.5 Correlation Matrix

	Hours	Mstatus	Age	Yrs_Ed
Hours	1			
Mstatus	−0.502	1		
Age	0.321	0.185	1	
Yrs_Ed	−0.588	0.040	−0.472	1

SOURCE: Authors' own.

Chapter 12 Correlation and Regression 211

Figure 12.7 Regression Modelling in Excel

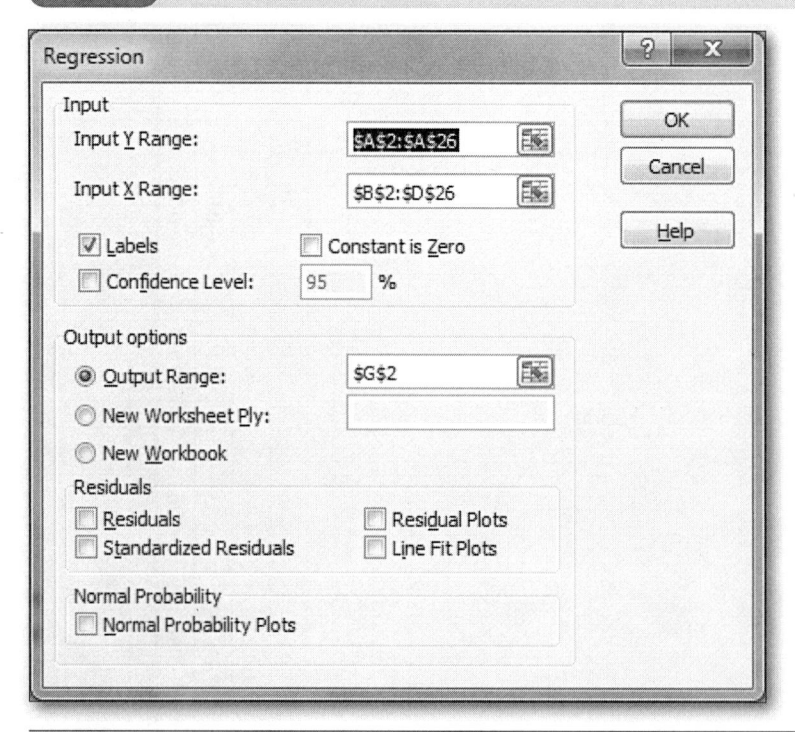

SOURCE: Authors' own.

Figure 12.8 Normal Probability Plot of the Residuals

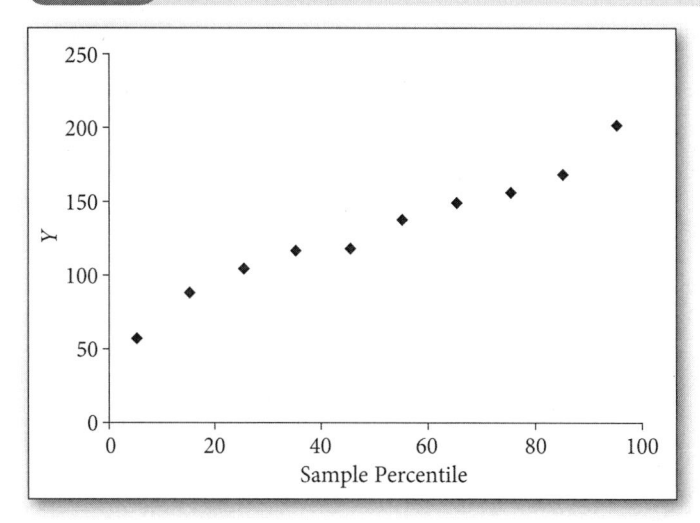

SOURCE: Authors' own.

Table 12.6 Summary Regression Output

Summary output

Regression Statistics	
Multiple R	0.776
R^2	0.602
Adjusted R^2	0.542
Standard error	0.772
Observations	24

ANOVA

	df	SS	MS	F	Significance F
Regression	3	18.006	6.002	10.065	0.000
Residual	20	11.927	0.596		
Total	23	29.933			

	Coefficients	Standard Error	t-Stat	P-value	Lower 95%	Upper 95%	Lower 95.0%	Upper 95.0%
Intercept	1.502	2.716	0.553	0.586	-4.164	7.168	-4.164	7.168
Mstatus	-1.174	0.329	-3.569	0.002	-1.860	-0.488	-1.860	-0.488
Age	0.039	0.033	1.165	0.258	-0.031	0.108	-0.031	0.108
Yrs_Ed	-0.152	0.052	-2.942	0.008	-0.260	-0.044	-0.260	-0.044

SOURCE: Authors' own.

Table 12.7 Summary Regression Output

Summary Output

Regression Statistics	
Multiple R	0.774
R^2	0.599
Adjusted R^2	0.563
Standard error	0.762
Observations	25

ANOVA

	df	SS	MS	F	Significance F
Regression	2	19.096	9.548	16.456	0.000
Residual	22	12.764	0.580		
Total	24	31.86			

	Coefficients	Standard Error	t-Stat	P-value	Lower 95%	Upper 95%
Intercept	4.646	0.473	9.823	0.000	3.665	5.627
Yrs_Ed	−0.184	0.043	−4.245	0.000	−0.274	−0.094
Mstatus	−1.096	0.312	−3.513	0.002	−1.743	−0.449

Thus the model is: Hours = 4.6 − 1.1 M status − 0.18Yrs_Ed

SOURCE: Authors' own.

12.6 MODELLING

This involves comparisons of different models. The most suitable model is selected on the following grounds:

1. Making theoretical sense
2. Adequacy of fit
3. Explanatory power
4. Parsimony

Figure 12.9 Diagnostic Plots

Yrs_Ed Residual Plot

Mstatus Residual Plot

Normal Probability Plot

SOURCE: Authors' own.

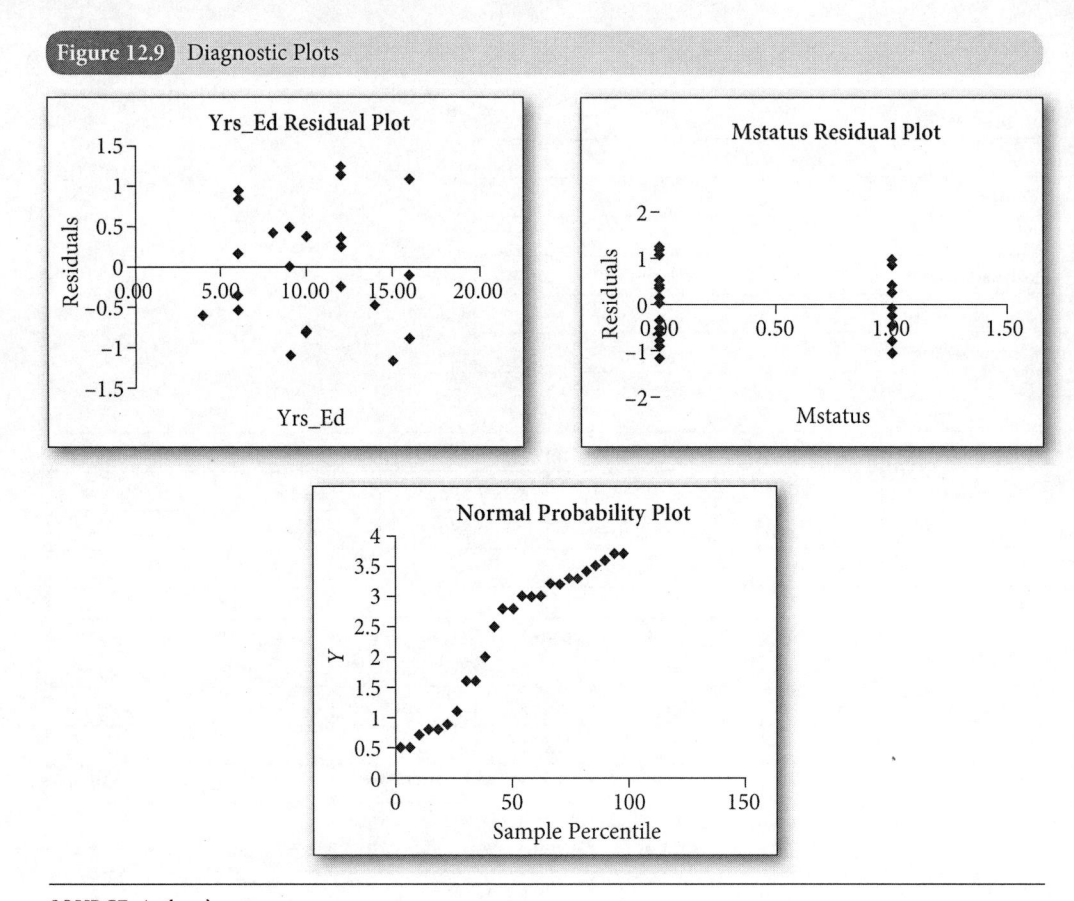

Example 1

Consider the case of modelling the miles per gallon of fuel used by cars—we might have explanatory variables of 0–60 acceleration, engine displacement, number of cylinders, weight of the car and year of manufacture.

This gives the model displayed in Table 12.8.

So only weight and model year appear as significant yet the Adjusted R^2 is reasonable. It is surprising that acceleration is not important; however, entering acceleration on its own gives acceleration significance. This is typical of the effects of multicolinearity, which arises when the so-called independent variables are highly correlated. This can result in good fitting models in which the coefficients appear to be insignificant and sometimes their signs are in unexpected directions. To test for multicolinearity examine the correlation matrix (see Table 12.9).

Table 12.8 Regression Model of Fuel Used by Cars

Regression Statistics	
Multiple R	0.899
R^2	0.808
Adjusted R^2	0.805
Standard error	3.440
Observations	390

ANOVA

	df	SS	MS	F	Sig F
Regression	6	19,048	3,175	268	0
Residual	383	4,531	12		
Total	389	23,580			

	Coefficients	SE	t-Stat	P-value	Lower 95%	Upper 95%
Intercept	−14.655	4.803	−3.051	0.002	−24.099	−5.211
Engine displacement	0.008	0.007	1.125	0.261	−0.006	0.023
Horsepower	−0.001	0.014	−0.087	0.931	−0.029	0.027
Vehicle weights	−0.007	0.001	−10.12	0.000	−0.008	−0.005
Time to accelerate	0.080	0.104	0.771	0.441	−0.125	0.285
Model year	0.758	0.053	14.33	0.000	0.654	0.862
Number of cylinders	−0.367	0.335	−1.097	0.274	−1.025	0.291

SOURCE: Authors' own.

Table 12.9 Correlation Matrix for Car Example

	Miles	Engine	Horse Power	Weight	Accelerate	Model Year	Cylinders
Miles per gallon	1						
Engine	−0.805	1					
Horse power	−0.776	0.898	1				
Weight	−0.831	0.934	0.863	1			
Accelerate	0.431	−0.548	−0.701	−0.425	1		
Model year	0.577	−0.367	−0.411	−0.303	0.296	1	
Cylinders	−0.776	0.951	0.842	0.897	−0.511	−0.342	1

SOURCE: Authors' own.

Hence, the independent variables are not independent as the coefficients are fairly large in magnitude. To deal with this do not include the correlated variable(s) in the model. Another approach is to combine the highly correlated variables together to form a new variable. This is done by a procedure called Factor Analysis (or Principal Components)—this is explained in the following chapter.

In this example, weight is highly correlated with every other variable—so using only weight provides a model that fits quite well but only includes a single independent variable rather than six independent variables in the full model.

The R^2 is 69 per cent, which is not too bad as there is only one independent variable in the model. However, when the diagnostic plots are examined, another problem appears (see Table 12.10).

The 'U'-shaped pattern in the first plot shown in Figure 12.10 and deviance from a straight line shown in the second plot in Figure 12.10 suggest that the relationship is not a straight-line one but may be a curve. A model of the form $y = \beta_0 + \beta_1 x_1 + \beta_2 x_1^2 + \varepsilon$ might be a solution (this is a quadratic model).

Fitting this model gives an adjusted R^2 of 65.5 per cent and the estimates of the coefficients are presented in Table 12.11.

However, from Figure 12.11 one can observe that although the 'U' shape is not so apparent the residuals 'fan out' and an outlier appear. This situation of increasing

Table 12.10 Regression with Only Weight as an Independent Variable

Summary Output

Regression Statistics	
Multiple R	0.831
R^2	0.690
Adjusted R^2	0.689
Standard error	4.338
Observations	391

ANOVA

	df	SS	MS	F	Sig F
Regression	1	16,290	16,290	865.645	5.884E-101
Residual	389	7,320	19		
Total	390	23,610			

	Coefficients	Standard Error	t-Stat	P-value
Intercept	46.200	0.803	57.557	0.000
Vehicle weight (lbs)	−0.008	0.000	−29.422	0.000

SOURCE: Authors' own.

Figure 12.10 Diagnostic Plots of Residuals for the Model Fitting Fuel Use by Weight Alone

SOURCE: Authors' own.

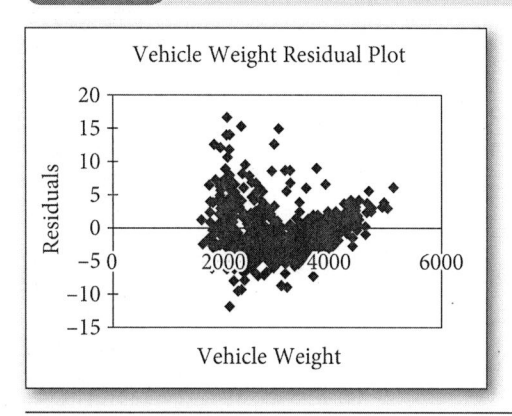

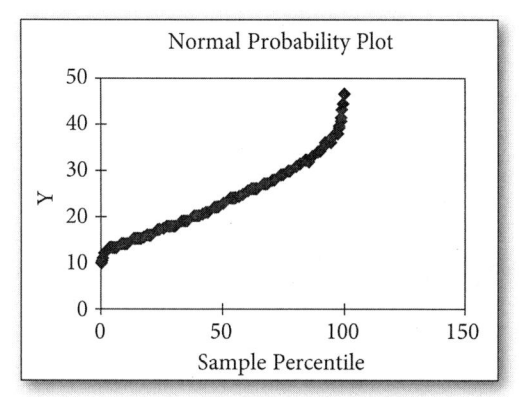

Table 12.11 Estimates of Coefficients

Model		Unstandardised Coefficients		Standardised Coefficients	t	Sig. Std. Error
		B	Std. Error	β	B	
1	(Constant)	52.540	3.030		17.337	.000
	Vehicle weight (lbs)	−.012	.002	−1.330	−6.094	.000
	Weight^2	7.60E-007	.000	.528	2.419	.016

SOURCE: Authors' own.
NOTE: Dependent variable: Miles per Gallon.

Figure 12.11 Residuals from the Quadratic Model of Fuel Use

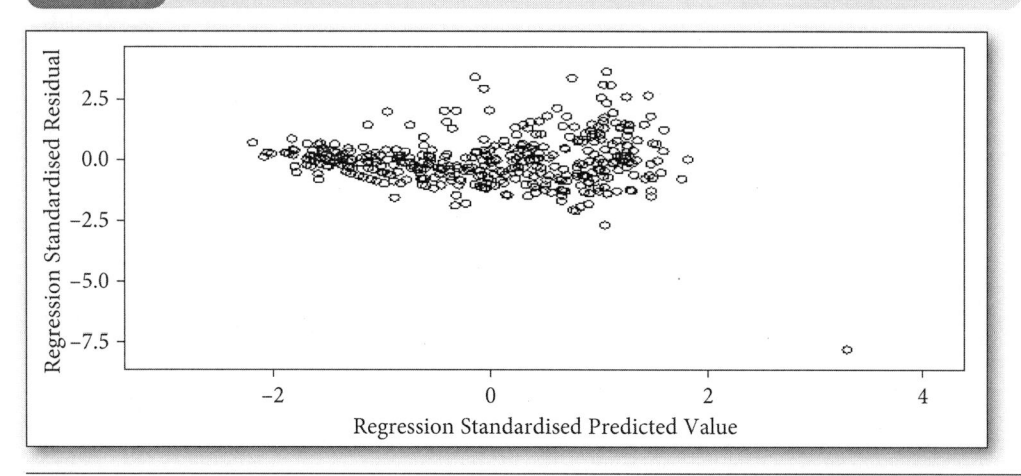

SOURCE: Authors' own.

variance in the residuals with higher values of the predicted values is frequently encountered and is known as *heteroscedasticity*. To correct for this, take logs of the dependent variable. The outlier was found to be case number 35 and has been removed from the data set—assuming that there has been some measurement error. The adjusted R^2 now rises to 76.7 per cent and the new estimates are displayed in Table 12.12.

The residuals from this model now appear acceptable and to conform to assumptions of normality as can be observed from Figure 12.12.

Table 12.12 Model of Log of Fuel Used

Model		Unstandardised Coefficients		Standardised Coefficients	t	Sig. Std. Error
		B	Std. Error	β	B	
1	(Constant)	4.416	.116		38.040	.000
	Vehicle Weight (lbs)	−.001	.000	−1.342	−6.998	.000
	Weight^2	2.93E-008	.000	.471	2.459	.014

SOURCE: Authors' own.
NOTE: Dependent variable: ln_fuel.

Figure 12.12 Residuals from the Log of Fuel Used Model

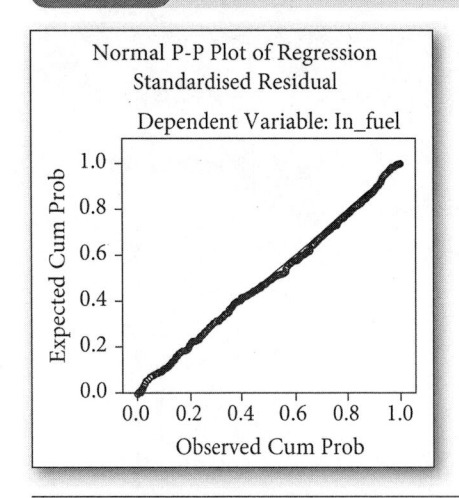

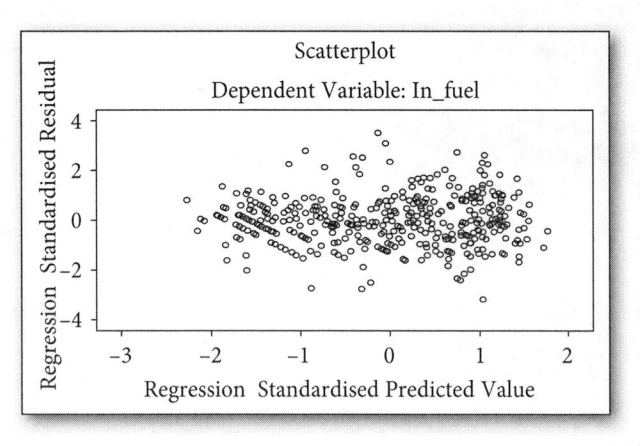

SOURCE: Authors' own.

12.7 EXERCISES

The data that will be used in these examples is available on the website http://www.sagepub.in/books/Book242499/samples
Load the file SIMREG.XLS (http://www.sagepub.in/books/Book242499/samples)

1. Determine a model of the cost of stay in hospital (cost), which is dependent on the duration of stay in days (day).
   Obtain the scatter plot—are any adjustments required?
   Correlation coefficient =
   R^2 =
   Model =
   Standard errors of coefficients =
   Obtain the residual plot and test residuals for normality.
   Predict how much an eight-day and a twenty-day stay will cost—which will be the most reliable?

2. Determine a model of how sales volume (svol) depends on the number of contacts a salesperson has. Try a model of the form $Y = a + bX$ and one of the form $Y = a + bX + cX^2$.
   For $Y = a + bX$
   R^2 =
   Model =
   Standard Errors of Coefficients =
   For $Y = a + bX + cX^2$
   R^2 =
   Model =
   Standard errors of coefficients =
   Which is the best model?
   What do you conclude about the plausibility of this model over the long run?

3. A company reckons that sales are exponentially related to advertising expenditure. Use the data advertex and sales to test this hypothesis by fitting the model:
   sales =
   Note that in Excel the exponential function is written as = exp(cell ref)
   This can be re-expressed as
   $\ln(\text{sales}) = \beta_0 + \beta_1\, advertex$
   R^2 =
   Model =
   Standard errors of coefficients =

4. Is there a relationship between heart attacks and wine consumption—use the variables wine and heart.
   Correlation coefficient =

$R^2 =$
Model =
Standard errors of coefficients =

5. Determine how demand for electricity (load) in California depends on temperature (temp).
   Correlation coefficient =
   $R^2 =$
   Model =
   Standard errors of coefficients =

6. The variables pe-ratio, profit, growth and employ are the price earnings ratio for 19 companies (firm) and their associated profit margins, percentage growth rates and number of employees in thousands. Obtain and interpret a multiple regression model with pe-ratio as the dependent variable.
   Correlation coefficient =
   $R^2 =$
   Model =
   Standard errors of coefficients =

7. The file Telser.xls contains data on the demand for cigarettes—use this data to obtain a model of cigarette demand. This data was extracted from Telser (1962). Locate file from http://www.sagepub.in/books/Book242499/samples.

8. The following is a table of quarterly sales for a company for the last four years. Use a regression model to predict the sales in 1998.

Year	Sales			
	Quarter 1	Quarter 2	Quarter 3	Quarter 4
1994	1,000	800	550	1,600
1995	1,105	890	630	1,808
1996	1,180	970	780	1,950
1997	2,050	1,020	830	

To do this make use of dummy variables to represent the quarter, that is create a variable D1 that will have the value 1 if it refers to quarter 1 and zero otherwise. Similarly create variables D2 and D3 to represent quarters 2 and 3. We do not need a variable for quarter 4 because if D1, D2 and D3 are all 0, then this implies quarter 4.

12.8 REFERENCE

L.G. Telser, 'Advertising and Cigarettes', *The Journal of Political Economy*, 70, no. 5 (1962): 471–499.

CHAPTER 13

Advanced Statistical Analysis

13.1 INTRODUCTION

In this chapter, we shall consider two of the most widely used advanced statistical methods: factor analysis and logistic regression. (There are several other useful techniques, but these ones are widely applied.) Factor analysis is widely used in business research to reflect hidden or latent variables which cannot be directly measured, but tend to be indirectly measured by other measures such as a bank or series of questions. Some examples are intelligence quotient, ambition, commitment and technical prowess.

Logistic regression is used when the outcome variable is binary or dichotomous, for example, success or failure, good credit risk or bad credit risk. In effect, there are just two outcomes. This technique has become very widely used in business decision-making, especially in the financial sector.

Owing to the complex nature of these techniques, one has to use a specialist computer package such as SPSS, as Excel just does not have the functionality to handle complex multivariate tasks. Here, SPSS will be used to outline these techniques and the datasets used in the examples can be downloaded from the web page associated with the book. Freeware software is available called R. The R project seeks to make statistical software available for everyone and this project has made available excellent and highly versatile software routines. For more information see http://www.r-project.org/ and also the book *Understanding Statistics using R* by Field et al. (2012).

This chapter is only vital for those who will be doing more advance analysis, probably based on survey data, and as such may be left unread by many. A good book to accompany this chapter is S. Sharma's *Applied Multivariate Techniques* (1996).

13.2 FACTOR ANALYSIS

In SPSS under data reduction one can employ Factor Analysis to reduce a given dataset to fewer variables. The objective is to form new variables by finding a linear combination of variables which are highly correlated. These new variables are created in a way so as to be orthogonal or uncorrelated.

This means that besides making the data more manageable, by reducing the number of variables, it is also a means to overcome the problems of multicolinearity. This procedure can also be used to create variables for phenomena that are not directly measurable such as intelligence quotient, love or empathy.

The method was discovered by Charles Spearman in 1904 while examining a correlation matrix of boys' test scores. This matrix is displayed in Table 13.1.

This suggested that the scores are related to some hidden factor, that is,

$$X_i = a_i F + e_i$$

where X_i is a standardised score, a_i is the factor loading and its square is the proportion of the variance accounted for by the factor. From this, Spearman developed a two-factor theory of mental tests: one common to all tests or general intelligence, and one specific to that test or specific intelligence. This was later modified to allow a test result to consist of several common factors plus a part specific to the test.

$a^2_{i1} + a^2_{i2} + \ldots + a^2_{im}$ are the communality of X_i, the part of its variance that is related to the common factors. $Var(e_i)$ is called the specificity of X_i. (This is the part that is unrelated to the common factors.)

So, Music = $0.8F + A_m$ and English could be $0.6F + A_e$.

The idea is to represent complex data by new variables formed by linear combinations of the data. These are called *factors* and one selects as many factors as needed to

Table 13.1 Correlation Matrix of Boys' Test Scores

	Classics	French	English	Mathematics	Discrimination of Pitch	Music
Classics	1.00	0.83	0.78	0.70	0.66	0.63
French		1.00	0.67	0.67	0.65	0.57
English			1.00	0.64	0.54	0.51
Mathematics				1.00	0.45	0.51
Discrimination of pitch						0.40
Music						1.00

SOURCE: Authors' own.

explain the variation in the data. One gets an idea from examining the *eigenvalues*. Often, eigenvalues greater than 1 are considered to be important; the rest are considered to be making a negligible contribution. To explain eigenvalues is beyond the scope of this chapter; those who are interested should consult Sharma (1996).

Stages for Factor Analysis

1. Construct the correlation matrix. Remove variables which are uncorrelated with others—they will not have a common factor.
2. Extract the factors—choose the required number of factors.
3. Rotate or transform the factors to make them more understandable.
4. Compute the scores for the factors.

The usual method for rotation is *Varimax* rotation. The idea here is that the interpretability of a factor can be measured by the variance of the square of its factor loadings.

Tests of Adequacy of the Approach

1. *Bartlett's test*. This is used to determine if values in the correlation matrix are different from an identity matrix. We want large significant values.
2. *Kaiser–Meyer–Olkin measure of sample adequacy* compares magnitudes of observed correlation coefficients to partial correlation coefficients. We want this to be large.

Kaiser gives values of

- KMO > 0.9 marvellous
- 0.8 < KMO < 0.9 meritorious
- 0.7 < KMO < 0.8 middling
- 0.6 < KMO < 0.7 mediocre
- 0.5 < KMO < 0.6 miserable
- KMO < 0.5 unacceptable

Example 1

A questionnaire was issued in an attempt to understand how German companies acquire knowledge. The file learning.xls is an Excel file and can be located at http://www.sagepub.in/books/Book242499/samples. This is an extract from a doctorate study by Lewrick (2007).

Some of the questions asked are summarised in Table 13.2.

Table 13.2 Questions in Summary Form—These Were Used to Ascertain How German Companies Acquire Knowledge

Question	Score Out of Five the Degree to Which Your Company Has Acquired Knowledge by the Following
Q1	From family & friends
Q2	From university
Q3	From business bodies
Q4	From informal contacts
Q5	From larger network
Q6	From experienced entrepreneurs
Q7	From markets via complementary products
Q8	From cooperative R&D
Q9	From introduction of new products/services
Q10	From design & manufacturing
Q11	Generated from a central position
Q12	People new to the company
Q13	From new introductions to the industry
Q14	From managerial & organisational skills
Q15	From staff training
Q16	From new innovation skills

SOURCE: Authors' own.

The correlation Matrix between the questions is shown in Table 13.3.

To activate factor analysis in SPSS—go to *Analyse*, then *Dimension Reduction* and *Factor*. The KMO and Bartlett tests are available under *descriptive* and one often wishes to *save* the factor scores, which are the new variables.

By activating factor analysis, the method was found to be suitable, giving a Kaiser–Meyer–Olkin measure of sampling adequacy of 0.813 and Bartlett's Test of sphericity is significant at the <0.001 level. Thus, the approach is sensible.

The procedure combines the questions to form new variables. The number of new variables formed depends on the amount of variation in the original questions, which is accounted for when new variables stop being constructed when there is little gain in the variance accounted for, usually detected when the eigenvalue falls below 1.

Table 13.4 shows the percentage of the original variation explained, and in this case four factors emerge.

So four factors are required and these explain 62.4 per cent of the original variation in the data. The factor loadings, that is, the share of the individual question scores, are tabulated in Table 13.5.

Table 13.3 Correlations between Questions

	Q1	Q2	Q3	Q4	Q5	Q6	Q7	Q8	Q9	Q10	Q11	Q12	Q13	Q14	Q15	Q16
Q1	1.000															
Q2	**0.276**	1.000														
Q3	0.101	**0.185**	1.000													
Q4	**0.387**	**0.430**	**0.342**	1.000												
Q5	−0.018	**0.292**	**0.420**	**0.243**	1.000											
Q6	**0.174**	0.145	**0.208**	**0.351**	**0.398**	1.000										
Q7	0.145	**0.267**	0.029	**0.377**	0.037	**0.272**	1.000									
Q8	−0.186	0.083	−0.106	−0.055	**0.205**	**0.271**	**0.296**	1.000								
Q9	−0.148	0.112	0.118	**0.213**	**0.187**	**0.299**	**0.465**	**0.376**	1.000							
Q10	−0.103	−0.042	−0.061	0.091	0.153	**0.296**	**0.299**	**0.541**	**0.576**	1.000						
Q11	−0.079	**0.240**	**0.311**	**0.359**	**0.263**	**0.414**	**0.386**	**0.377**	**0.462**	**0.390**	1.000					
Q12	0.118	0.081	−0.126	0.095	0.037	**0.178**	**0.333**	**0.166**	0.149	**0.370**	**0.179**	1.000				
Q13	−0.107	0.097	−0.106	−0.085	**0.178**	**0.170**	**0.219**	**0.401**	**0.339**	**0.385**	**0.336**	**0.538**	1.000			
Q14	−0.001	**0.415**	**0.156**	**0.378**	**0.308**	**0.270**	**0.270**	0.051	**0.330**	**0.260**	**0.418**	**0.236**	**0.196**	1.000		
Q15	0.074	**0.306**	**0.196**	**0.295**	**0.486**	**0.217**	**0.268**	**0.191**	**0.210**	**0.190**	**0.397**	**0.379**	**0.347**	**0.619**	1.000	
Q16	0.040	**0.265**	0.051	**0.193**	**0.388**	**0.362**	**0.205**	**0.346**	**0.300**	**0.294**	**0.441**	**0.510**	**0.489**	**0.445**	**0.580**	1.000

SOURCE: Authors' own.
NOTE: Figures in bold represent correlations which are significant at the 5 per cent level.

Table 13.4 Variance Explained by the Factors Created

Question	Eigen Value	% of Variance	Cumulative %	Total	% of Variance	Cumulative %	Total	% of Variance	Cumulative %
Q1	4.876	30.476	30.476	4.876	30.476	30.476	2.930	18.315	18.315
Q2	2.273	14.204	44.680	2.273	14.204	44.680	2.633	16.455	34.769
Q3	1.457	9.105	53.786	1.457	9.105	53.786	2.361	14.759	49.528
Q4	1.373	8.579	62.364	1.373	8.579	62.364	2.054	12.836	62.364
Q5	0.993	6.204	68.568						
Q6	0.812	5.075	73.643						
Q7	0.698	4.360	78.004						
Q8	0.605	3.783	81.787						
Q9	0.548	3.427	85.215						
Q10	0.521	3.259	88.474						
Q11	0.435	2.716	91.190						
Q12	0.396	2.476	93.666						
Q13	0.369	2.306	95.972						
Q14	0.271	1.691	97.663						
Q15	0.212	1.328	98.990						
Q16	0.162	1.010	100.000						

SOURCE: Authors' own.

Table 13.5 Component Matrix Showing Factor Loadings

Question	Factors			
	1	2	3	4
Q1	0.726	−0.093	0.286	−0.299
Q2	0.718	0.007	−0.300	0.046
Q3	0.693	0.171	0.262	−0.377
Q4	0.652	0.242	0.112	−0.165
Q5	0.617	−0.257	−0.379	0.268
Q6	0.578	−0.487	−0.199	0.219
Q7	0.572	0.141	−0.220	0.121
Q8	0.555	−0.488	0.273	−0.171
Q9	0.523	0.291	−0.247	−0.510
Q10	0.476	0.612	−0.020	0.360
Q11	0.249	0.559	−0.436	−0.195
Q12	0.072	0.513	0.422	0.421
Q13	0.500	−0.507	−0.245	0.052
Q14	0.435	0.485	0.190	0.051
Q15	0.508	−0.293	0.609	0.052
Q16	0.561	−0.020	0.043	0.578

SOURCE: Authors' own.

Table 13.5 is hard to interpret and the components are rescaled to maximise question loadings on the different factors—this is the rotated solution, which is generated by applying varimax rotation. The rotated solution is presented in Table 13.6.

This can be made similar to interpret if one clicks on options in the factor menu and sort coefficients by size and suppress all small coefficients, which are small associations. (We usually use 0.3 as a threshold value.)

This produces the output shown in Table 13.7.

We can now give the factors names, factor 1 can be called industry learning, factor 2 is given the name organisational network, factor 3 can be named external network and factor 4 is labelled individual network. These factors can be used as variables in the data set and we can now work with four variables rather than the 16 original questions. In addition, the factors created are uncorrelated or orthogonal; thus, they can be used in regression. Factor analysis can frequently be used as a solution to multicolinearity, which arises in regression when so-called independent variables are correlated.

SOURCE: Authors' own.

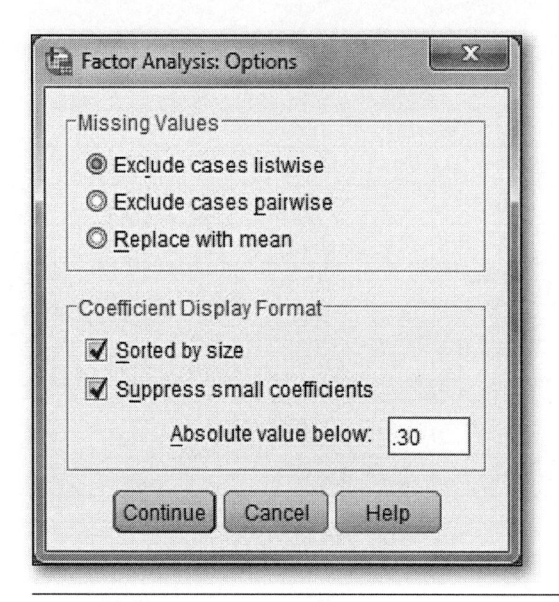

Table 13.6 Rotated Component Matrix

Questions	Factors			
	1	2	3	4
From introduction of new products/services	0.796	0.071	0.151	0.047
From design & manufacturing	0.766	0.252	−0.046	−0.082
From cooperative R&D	0.675	0.242	0.020	−0.236
Generated from a central position	0.607	0.188	0.430	0.137
From markets via complementary products	0.591	0.149	−0.098	0.520
From experienced entrepreneurs	0.452	0.095	0.363	0.256
People new to the company	0.180	0.769	−0.232	0.199
From new innovation skills	0.228	0.745	0.311	0.064
From new introductions to the industry	0.358	0.701	−0.049	−0.166
From staff training	0.068	0.664	0.496	0.170
From larger network	0.092	0.234	0.782	−0.059
From business bodies	0.029	−0.240	0.719	0.164
From managerial & organisational skills	0.174	0.442	0.466	0.282
From family & friends	−0.198	0.035	−0.099	0.758
From informal contacts	0.193	−0.040	0.363	0.750
From university	−0.011	0.226	0.352	0.537

SOURCE: Authors' own.

Table 13.7 Simplified Rotated Component Matrix

Questions	Component			
	1	2	3	4
From introduction of new products/services	.796			
From design & manufacturing	.766			
From cooperative R&D	.675			
Generated from a central position	.607		.430	
From markets via complementary products	.591			.520
From experienced entrepreneurs	.452		.363	
People new to the company		.769		
From new innovation skills		.745	.311	
From new introductions to the industry	.358	.701		
From staff training		.664	.496	
From larger network			.782	
From business bodies			.719	
From managerial & organisational skills		.442	.466	
From managerial & organisational skills				.758
From staff training			.363	.750
From new innovation skills			.352	.537

SOURCE: Authors' own.

Residuals are used to determine if the method is reliable—but it is the residual correlation matrix that is examined and low values are hoped for. This is found by subtracting the observed correlation matrix from a predicted correlation matrix. (This is obtained by going to *descriptives* and then *reproduced* in the correlation matrix box.) This is displayed in Table 13.8 and it is hoped that the reproduced correlation matrix is similar to the original correlation matrix of correlation between questions and that the residual correlations (original matrix—reproduced matrix) are small.

Hence, this looks acceptable. Thus, 16 variables have been reduced to four variables with an information loss of only 37.6 per cent.

Example 2

Consider the data in TFRxls (http://www.sagepub.in/books/Book242499/samples), which is a file extracted from the United Nations Development Program report on human development. It contains data on the total fertility rate (TFR, the average number

Table 13.8 Reproduced Correlations

		People New to the Company	From new Introductions to the Industry	From Managerial & Organisational Skills	From Staff Training	From New Innovation Skills	From Family & Friends	From University	From Business Bodies
Reproduced correlation	People new to the company	.717[a]	.582	.320	.441	.554	.165	.197	-.313
	From new introductions to the industry	.582	.650[a]	.303	.437	.578	-.167	.048	-.220
	From managerial & organisational skills	.320	.303	.523[a]	.584	.532	.149	.413	.281
	From staff training	.441	.437	.584	.720[a]	.675	.090	.415	.227
	From new innovation skills	.554	.578	.532	.675	.707[a]	-.001	.309	.062
	From family & friends	.165	-.167	.149	.090	-.001	.624[a]	.382	.039
	From university	.197	.048	.413	.415	.309	.382	.463[a]	.287
	From business bodies	-.313	-.220	.281	.227	.062	.039	.287	.603[a]
	From informal contacts	.070	-.101	.396	.294	.175	.493	.519	.399
	From larger network	.004	.168	.468	.539	.434	-.132	.296	.499
	From experienced entrepreneurs	.122	.168	.362	.317	.303	.072	.281	.293
	From markets via complementary products	.347	.234	.270	.179	.249	.293	.272	-.003
	From cooperative R&D	.256	.449	.167	.176	.325	-.306	-.073	-.063
	From introduction of new products/services	.172	.319	.253	.183	.284	-.134	.085	.122
	From design & manufacturing	.326	.467	.201	.182	.343	-.200	-.012	-.085
	Generated from a central position	.182	.305	.428	.403	.421	-.052	.261	.304

Residual[b]									
	People new to the company		-.044	-.084	-.062	-.044	-.047	-.116	.187
	From new introductions to the industry	-.044		-.107	-.090	-.088	.060	.049	.114
	From managerial & organisational skills	-.084	-.107		.034	-.087	-.149	.002	-.125
	From staff training	-.062	-.090	.034		-.095	-.015	-.108	-.032
	From new innovation skills	-.044	-.088	-.087	-.095		.041	-.044	-.011
	From family & friends	-.047	.060	-.149	-.015	.041		-.106	.062
	From university	-.116	.049	.002	-.108	-.044	-.106		-.102
	From business bodies	.187	.114	-.125	-.032	-.011	.062	-.102	
	From informal contacts	.026	.016	-.019	.002	.017	-.105	-.089	-.058
	From larger network	.033	.010	-.160	-.054	-.046	.113	-.003	-.080
	From experienced entrepreneurs	.056	.002	-.092	-.101	.059	.102	-.136	-.085
	From markets via complementary products	-.015	-.015	.000	.089	-.043	-.148	-.004	.032
	From cooperative R&D	-.090	-.048	-.117	.015	.021	.119	.155	-.043
	From introduction of new products/services	-.023	.020	.077	.027	.016	-.014	.027	-.005
	From design & manufacturing	.044	-.081	.059	.008	-.049	.096	-.031	.024
	Generated from a central position	-.004	.031	-.010	-.005	.021	-.027	-.020	.007

(Table 13.8 Contd)

(Tale 13.8 Contd)

		From Informal Contacts	From Larger Network	From Experienced Entrepreneurs	From Markets via Complementary Products	From Cooperative R&D	From Introduction of New Products/Services	From Design & Manufacturing	Generated from a Central Position
Reproduced correlation	People new to the company	.070	.004	.122	.347	.256	.172	.326	.182
	From new introductions to the industry	-.101	.168	.168	.234	.449	.319	.467	.305
	From managerial & organisational skills	.396	.468	.362	.270	.167	.253	.201	.428
	From staff training	.294	.539	.317	.179	.176	.183	.182	.403
	From new innovation skills	.175	.434	.303	.249	.325	.284	.343	.421
	From family & friends	.493	-.132	.072	.293	-.306	-.134	-.200	-.052
	From university	.519	.296	.281	.272	-.073	.085	-.012	.261
	From business bodies	.399	.499	.293	-.003	-.063	.122	-.085	.304
	From informal contacts	.732[a]	.248	.407	.463	-.049	.240	.060	.368
	From larger network	.248	.678[a]	.333	-.018	.148	.205	.098	.428
	From experienced entrepreneurs	.407	.333	.411[a]	.379	.275	.433	.333	.483
	From markets via complementary products	.463	-.018	.379	.652[a]	.310	.490	.452	.416
	From cooperative R&D	-.049	.148	.275	.310	.570[a]	.546	.596	.431
	From introduction of new products/services	.240	.205	.433	.490	.546	.663[a]	.617	.568
	From design & manufacturing	.060	.098	.333	.452	.596	.617	.659[a]	.482
	Generated from a central position	.368	.428	.483	.416	.431	.568	.482	.608[a]

Residual[b]									
	People new to the company	.026	.033	.056	-.015	-.090	-.023	.044	-.004
	From new introductions to the industry	.016	.010	.002	-.015	-.048	.020	-.081	.031
	From managerial & organisational skills	-.019	-.160	-.092	.000	-.117	.077	.059	-.010
	From staff training	.002	-.054	-.101	.089	.015	.027	.008	-.005
	From new innovation skills	.017	-.046	.059	-.043	.021	.016	-.049	.021
	From family & friends	-.105	.113	.102	-.148	.119	-.014	.096	-.027
	From university	-.089	-.003	-.136	-.004	.155	.027	-.031	-.020
	From business bodies	-.058	-.080	-.085	.032	-.043	-.005	.024	.007
	From informal contacts	-.005	-.005	-.056	-.085	-.005	-.027	.031	-.009
	From larger network	-.056	.066	.066	.055	.058	-.017	.055	-.165
	From experienced entrepreneurs			-.107	-.004	-.134	-.036	-.070	
	From markets via complementary products	-.085	.055	-.107		-.014	-.025	-.153	-.030
	From cooperative R&D	-.005	.058	-.004	-.014		-.170	-.056	-.054
	From introduction of new products/services	-.027	-.017	-.134	-.025	-.170		-.041	-.106
	From design & manufacturing	.031	.055	-.036	-.153	-.056	-.041		-.092
	Generated from a central position	-.009	-.165	-.070	-.030	-.054	-.106	-.092	

Extraction Method: Principal Component Analysis.
[a]Reproduced communalities.
[b]Residuals are computed between observed and reproduced correlations. There are 57 (47.0%) nonredundant residuals with absolute values greater than 0.05.

SOURCE: Authors' own.

of children a women has) and 'independent' variables GDP per capita, female life expectancy at birth (Fexp), female literacy (Flit) and infant mortality (Imort). As demographic theory has it, when countries get wealthier and female well-being and education increase, fertility should fall. If infant mortality is high, then mothers have a desire to 'replace' children who have died or those that they anticipate will die. To test these suppositions, one can fit a multiple regression model. The model fits well with an adjusted R^2 of over 77 per cent but as can be seen from Table 13.9 something is wrong—gdp per capita of one of the theoretically most important variables is not significant. This is the effect of the 'independent' variables being highly correlated (see Table 13.10)—an effect called multicolinearity.

To overcome this, factor analysis can be applied and it seems suited because of the high Kaiser–Meyer–Olkin statistic of 0.777 and the Bartlett's test value being significant

Table 13.9 Multiple Regression Model to Explain Fertility

Variables	Unstandardised Coefficients		Standardised Coefficients	t	Sig.
	B	Std. Error	Beta		
(Constant)	6.521	1.272		5.125	.000
ln of GDP per capita	−.084	.088	−.065	−.954	.342
Female life expectancy 2004	−.019	0.012	−.157	−1.565	.121
Female adult literacy ratio	−.027	0.005	−.379	−5.555	.000
Infant mortality rate per 1,000 in 2004	.016	.005	.383	3.228	.002

SOURCE: Authors' own.

Table 13.10 Correlation Matrix

	Total Fertility Rate	ln of GDP Per Capita	Female Life Expectancy	Female Adult Literacy Ratio	Infant Mortality Rate per 1,000
Total fertility rate	1	−.730**	−.806**	−.792**	.845**
ln of GDP per capita		1	−.739**	.601**	−.770**
Female life expectancy			1	.618**	−.720**
Female adult literacy ratio				1	.601**
Infant mortality rate per 1,000					1

SOURCE: Authors' own.
**Signifies correlation coefficient is significant at the 5% level ($P = 0.05$).

Table 13.11 Total Variance Explained of Fertility Data

Component	Initial Eigenvalues			Extraction Sums of Squared Loadings		
	Total	% of Variance	Cumulative %	Total	% of Variance	Cumulative %
1	3.101	77.532	77.532	3.101	77.532	77.532
2	.417	10.420	87.951			
3	.383	9.570	97.521			
4	.099	2.479	100.000			

SOURCE: Authors' own.

Table 13.12 Component Matrix of Car Data

	Component
	1
ln of GDP per capita	.839
Female life expectancy 2004	.901
Female adult literacy ratio	.827
Infant mortality rate per 1,000 in 2004	−.949

SOURCE: Authors' own.

at the <0.001 level. One factor is generated, which explains 71.5 per cent of the variance as is displayed in Table 13.11.

When only one factor is derived, no rotated solution is computed. The component matrix is shown in Table 13.12.

The reproduced and residual correlations were found to be acceptable. Thus, the four variables thought to affect fuel consumption have been reduced to one. This new variable can be used in regression modelling to explain fertility and the simple model; TFR = 3.433 − 1.499 * Factor and the adjusted R^2 for this model are still good at 75 per cent. Factor here is akin to the value of the Human Development Index, which is computed by UNDP.

13.3 LOGISTIC REGRESSION

Ordinary multiple regression is used to examine influences on a continuous variable. Logistic regression is used when the outcome is a proportion. It is widely used in social, financial, marketing and medical statistics for such applications as credit-scoring and predicting disease severity and progression to inform health-care management.

If we used ordinary regression for proportions, we would be likely to hit trouble, because we might be predicting proportions less than zero or greater than one. To get over this, we do not predict the proportion itself, but the logarithm of the odds of the proportion. If the proportion is p then the *logit* function of p is just $log(p/(1 - p))$. As p goes from 0 to 1 the *logit* function goes between minus infinity and positive infinity. The *logit* corresponding to a proportion of 0.5 is zero. The shape of the *logit* function has a sigmoid shape as shown here in Figure 13.1.

The inverse of the *logit* transformation (L) is used to calculate back to a proportion from L. It is $exp(L)/(1+exp(L))$. To calculate a *logit* (L) from p or a proportion from L, on a calculator,

L from p

1. From the proportion (p) calculate $p/(1 - p)$
2. Use the ln key to get the Napier or natural log of this

p from L

1. Calculate the exponential of L, usually exp. or e^x
2. $exp(L)/(1+exp(L))$ gives you p

Table 13.13 gives some proportions and *logits*—complete this table using the above rules, and check your results against the graph in Figure 13.1.

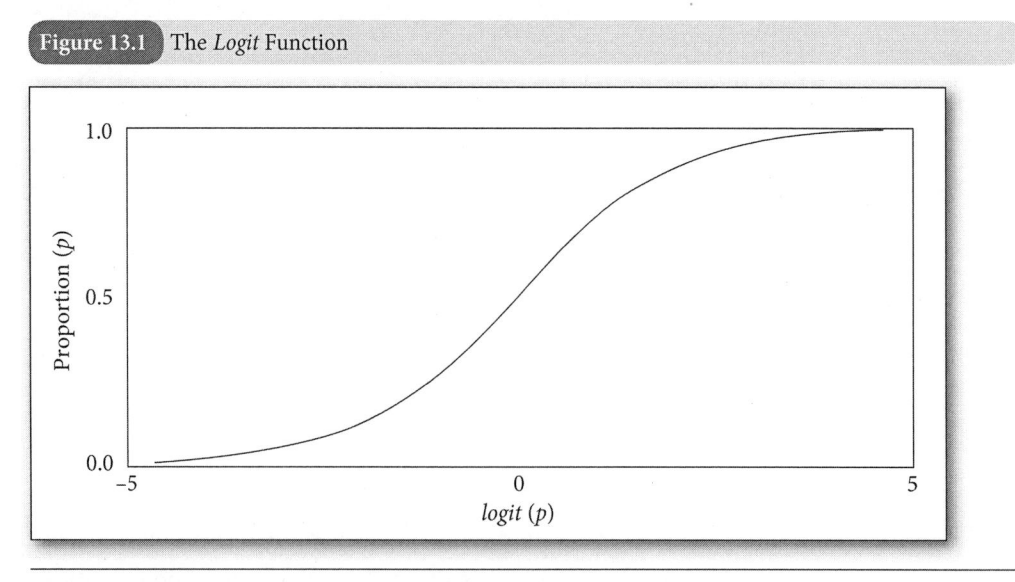

Figure 13.1 The *Logit* Function

SOURCE: Authors' own.

Table 13.13 Proportions and *Logits*

p	0.01	0.3	0.5	0.7	0.9	0.999
logit (p)	−4.60	−2.20	0.0	0.41	0.85	3.48

SOURCE: Authors' own.

Logistic regression is very similar to ordinary regression, but the Y variable is *logit (p)* instead of p itself. We can interpret the coefficients in logistic regression as an increase in the log-odds of the dependent variable, for each unit change of the x variable. When the coefficients are not too large, the increase in the log-odds can be interpreted as the proportional increase per unit change of x.

The coefficients and P-values have the same interpretation in logistic regression as in ordinary regression. To obtain logistic regression in SPSS follow the path:

Analyze > Regression > Binary Logistic.

The analysis below uses some data on first year students at Napier and on the factors that influence whether they pass their end of year exams at the first attempt. (The data is in the file *Napier.xls* [http://www.sagepub.in/books/Book242499/samples].)

In ordinary regression

- Start with scatter plots of all your possible x variables.

In logistic regression

- Start with tables of your y variable with each of your x variables, grouping the x's if necessary. Be sure to calculate the per cents for your dependent variable.

When all variables are included together in a logistic regression, the factors are adjusted for each other.

Doing lots of complicated tables would be an alternative to logistic regression. But the regression approach helps organise things and also allows you to get a score for the most important factors.

Example 2: Predicting Student Performance at Edinburgh Napier University

The data consists of all first year students entering the first year of the modular course in 1994–1995, with Highers as their qualifications at entry, and complete data on age and gender. The variable *outcome* describes what happened to them at the end of the year. It is coded as shown in Table 13.14.

Table 13.14 Variables Used in Student Progression Model

	Value Label	Frequency	Per cent	Valid Per cent	Cumulative Per cent
Withdrawn semester 1	1	88	6.5	6.5	6.5
Withdrawn semester 2	2	35	2.6	2.6	9.0
Fail	3	254	18.6	18.6	27.6
Pass resit	4	155	11.4	11.4	39.0
Pass 1st time	5	832	61.0	61.0	100.0
Total		1,364	100	100	

SOURCE: Authors' own.

Table 13.15 Progression and Gender

Gender	Don't Progress 0.00		Progress 1.00	
	Count	Row %	Count	Row %
Male	267	34.0	519	66.0
Female	110	19.0	468	81.0
Total	377	27.6	987	72.4

SOURCE: Authors' own.

The objective of the analysis is to get a score to predict progression. So we decide that we will consider values 4 and 5 as constituting progression, and recode them to get a new variable called *progress*, with value 1 if progressing and 0 otherwise. The data can be downloaded from http://www.sagepub.in/books/Book242499/samples.

The variables we have that could predict outcome are: gender, numbers of higher passes and age of student. Taking gender first we can look at Table 13.15.

We can see that there is a substantial difference between the pass rates for the two genders. This difference can be tested formally with a Chi-square test (see Chapter 11).

Now we can set this up as a prediction equation with logistic regression. Gender is coded as 1 for male and 2 for female. To make things easier, we will calculate a new dummy variable called *gendum* with the value of 0 for male and 1 for female. We will then fit a logistic regression with this as the explanatory variable.

The coefficients derived are shown in Table 13.16.

Thus, *logit* (progress) = 0.665 + 0.783 * gendum

The dummy variable for gender takes the value 0 for male and 1 for female, so the prediction equation for *logit* (pass rate) can be computed. Do this by completing Table 13.17; referring back to Table 13.13 will help you with this task.

Table 13.16 Variables in the Equation

Variable	B	S.E.	Wald	df	Sig.	R	Exp(B)
GENDUM	.783	.130	36.305	1	.000	.146	2.189
Constant	.665	.075	77.884	1	.000		

SOURCE: Authors' own.

Table 13.17 Predictions from the Logistic Regression Model

Prediction Equation	Predicted *logit*	Predicted Proportion
Males	constant + 0 × gendum	
Females	constant + 1 × gendum	

SOURCE: Authors' own.

Table 13.18 Success Rates of the Logistic Model

Observed Progress	Predicted Progress		
	No	Yes	Per cent Correct
No	0	377	0
Yes	0	987	100
Overall correct	72.36		

SOURCE: Authors' own.

The success of the model is given by the classification table for progress shown in Table 13.18. This is computed as part of the SPSS algorithm.

This may seem a long, roundabout way for doing something simple. The pay-off comes when we look at several factors together (below).

Multivariate Logistic Regression

We might now ask whether the greater success of female students might be attributed to their coming in with better qualifications, or because they had different ages at entry. We will now fit a logistic regression with all three dependent variables and the results are shown in Table 13.19.

(*Important note: If you were doing this on a real example you would take time to look at the individual explanatory variables by themselves first.*)

Table 13.19 Variables in the Equation

Variable	B	S.E.	Wald	df	Sig	R	Exp(B)
GENDUM	0.7655	0.1308	34.2675	1	0.0000	0.1416	2.1501
AGE	0.0394	0.0159	6.1579	1	0.0131	0.0508	1.0402
NHPASSES	0.1631	0.0509	10.2607	1	0.0014	0.0717	0.1771
Constant	-0.7415	0.4164	3.1701	1	0.0750		

SOURCE: Authors' own.

Here is the new prediction equation. We can see that all the variables are significant predictors of progression.

Interpretation of the coefficients for the independent variables:

GENDUM: _____
AGE: _____
NHPASSES: _____

Now calculate the predicted probability of progression for:

- A 17-year-old male student with 1 pass in highers.
- A 20-year-old female student with 4 passes in highers.

Assessing the Quality of the Models

R-squared is not available for logistic regression; however, guidance is available from the proportion correctly classified. The overall quality of the model is assessed by the *Scaled Deviance* (D).

This has a Chi-squared distribution with n-p degrees of freedom. This measure is particularly useful for comparing models in that models are deemed to be significantly different if $D_1 - D_2$ is greater than the Chi-squared distribution with q degrees of freedom. This is used in model building and a useful equivalence to forward selection is *forward log ratio* (LR). One must also think about the logic of the coefficients and their significance.

Example 4

In the file *credit.xls* there is a variable credit rating, which classifies bank customers into good and bad risks. It is thought that this binary variable might be predicted by the customers' age, whether or not they have a visa card, whether or not the customer has a mortgage and their occupation type (managerial, professional, clerical, skilled manual or unskilled). Binary logistic regression is applied using the menu:

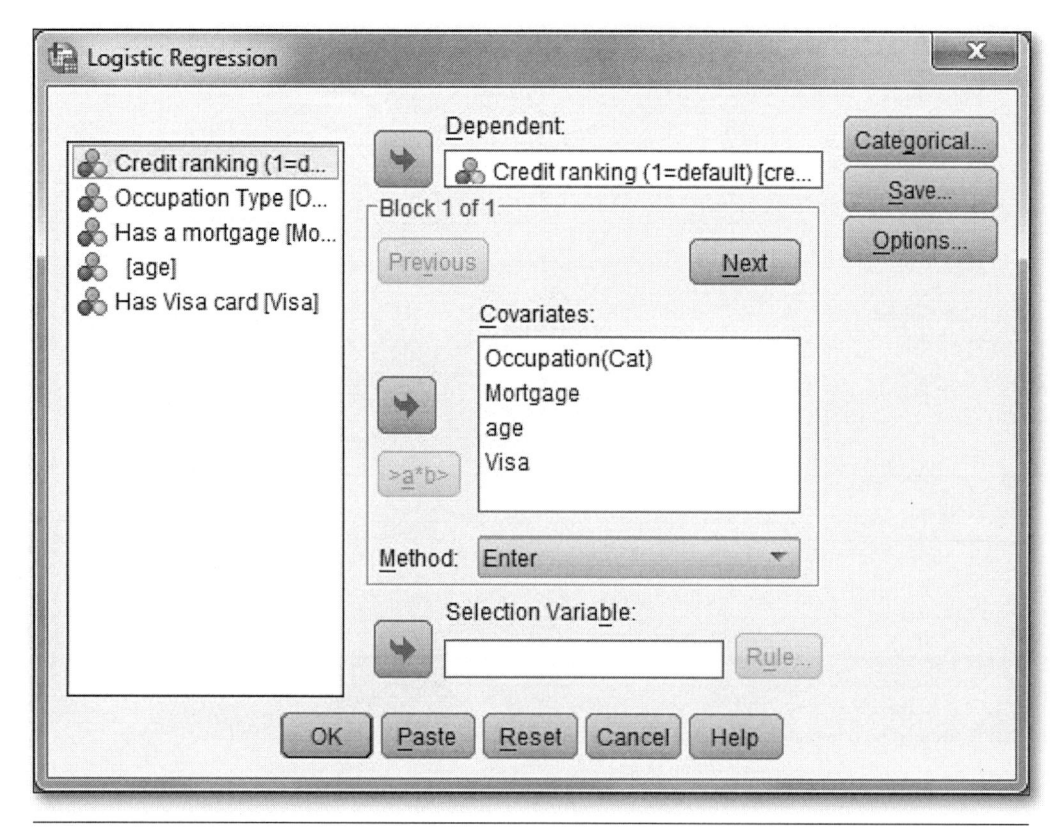

SOURCE: Authors' own.

Occupation type is categorical and is set via the categorical button into a series of dummy variables as detailed in Table 13.20.

In this categorisation unskilled is zero in all columns; this is the baseline that other occupation categories are compared to.

The model fits well as can be determined from the fit metrics of Cox and Snell and Nagelkerke shown in Table 13.21, and how well good and bad credit is predicted, see Table 13.22.

This table is compared to the performance of the null model (a model without any variables); this model only had a predictive ability of 52 per cent, so our model is considerably better.

The model is given in Table 13.23.

Table 13.20 Categorisation of Occupation Type

		Categorical Variables' Codings				
		Frequency	Parameter Coding			
			(1)	(2)	(3)	(4)
Occupation type	Management	39	1.000	.000	.000	.000
	Professional	158	.000	1.000	.000	.000
	Clerical	47	.000	.000	1.000	.000
	Skilled manual	41	.000	.000	.000	1.000
	Unskilled	38	.000	.000	.000	.000

SOURCE: Authors' own.

Table 13.21 Fit Metrics

	Model Summary		
Step	−2 Log likelihood	Cox & Snell R^2	Nagelkerke R^2
1	191.295[a]	.547	.730

SOURCE: Authors' own.
NOTE: [a]Estimation terminated at iteration number 7 because parameter estimates changed by less than .001.

Table 13.22 Predictive Performance of the Model

Observed		Predicted		
		Credit Ranking (1=default)		Percentage Correct
		Bad	Good	
Credit ranking (1=default)	Bad	146	22	86.9
	Good	17	138	89.0
Overall percentage				87.9

SOURCE: Authors' own.

Table 13.23 Coefficients of the Model to Predict Credit Worthiness

		Variables in the Equation					
		B	S.E.	Wald	df	Sig	Exp(B)
Step 1[a]	Occupation			8.748	4	0.68	0
	Occupation (1)	3.225	1.394	5.355	1	0.21	25.166
	Occupation (2)	−.233	.687	.115	1	.734	.792
	Occupation (3)	−.177	.747	.056	1	.813	.838
	Occupation (4)	−1.436	1.158	1.538	1	.215	.238
	Mortgage	3.170	.482	43.278	1	.000	23.811
	age	.142	.027	28.530	1	.000	1.152
	Visa	−.314	.383	.673	1	.412	.731
	Constant	−5.656	0.942	36.086	1	.000	.003

SOURCE: Authors' own.
NOTE: [a] Variable(s) entered on step 1: Occupation, Mortgage, age, Visa.

Here the *logit*(credit rating) = ln(Prob(good risk)/(1 − Prob(good risk))) = −5.656 − 0.324 * Visa + 0.142 * age + 3.17 * mortgage − 1.436 * Semi Skilled − 0.177 * Clerical − 0.233 * Professional + 3.225 * managerial

From this model, one can conclude that credit worthiness is higher for managers and older people. Having a mortgage is also associated with credit worthiness. All other occupational types are not significantly different from unskilled workers and having Visa card does not significantly affect the propensity to be credit worthy.

13.4 EXERCISES

1. The file Contract.xls (obtain the data from http://www.sagepub.in/books/Book242499/samples) contains data on the degree to which an outsourcing arrangement was successful and how the company scored on a rating questionnaire prior to establishing the relationship. Use factor analysis to construct some latent variables and develop a model to allow the success rating to be estimated from the questionnaire scores.

   The questionnaire was: Rate the potential outsourcer on the scales shown in Table 13.24.

Table 13.24 Criteria for Selecting Outsource Partner

Criteria	Very Weak				Very Strong
	1	2	3	4	5
Technical capability					
Ability to manage projects					
Collaborator's track record					
Strength of communication links					
Compatibility of operating cultures					
Compatibility of strategic aims					
Collaborator's development speed					
Collaborators development costs					
Business strength of collaborator					
Financial resources of collaborator					
Security					
Strategic position					

SOURCE: Authors' own.
NOTE: This data was collected by William Bailiey as part of his PhD (1997).

2. Use the file learning.xls used in Section 13.2 to compute a logistic model to determine how each of the derived factors contribute towards success.

13.5 REFERENCES

W.J. Bailey, 'The Selection of Collaborative Technology Development Partners' (PhD thesis, Edinburgh Napier University, Edinburgh, 1997).
J. Bongaarts and S.C. Watkins, 'Social Interactions and Contemporary Fertility Transitions', *Population and Development Review* 22 (1996): 639–682.
A. Field, J. Miles and Z. Field, *Understanding Statistics Using R* (London: SAGE Publications, 2012).
M. Lewrick, 'Changes in Innovation Styles' (PhD thesis, Edinbuurgh Napier University, Edinburgh, 2007).
R Project, http://www.r-project.org/ (accessed December 2012).
S. Sharma, *Applied Multivariate Techniques* (Chichester: Wiley, 1996).
C. Spearman, 'General Intelligence, Objectively Determined and Measured', *American Journal of Psychology*, 15 (1904): 201–293.

CHAPTER 14

Tests of Measurement and Quality

14.1 INTRODUCTION

Three criteria are generally used for testing and evaluating measurements of variables and ensuring the quality of data, research design methods and the overall accuracy of study results. They are known to be reliability, validity and generalisability. These are very important both in qualitative and in quantitative research. In quantitative research they are followed fairly easily in order to understand the actual reality and to generalise the findings. However, qualitative study has different benefits although generalisation is hard. Qualitative research requires theoretical sophistication and methodological rigour.

14.2 RELIABILITY

Reliability estimates the consistency of the measurement or more simply the degree to which an instrument measures the same way each time it is used under the same conditions with the same subjects. This is essentially about consistency. That is, if we measure something many times and the result is always the same then we can say that our measurement instrument is reliable. In other words, when the outcome of the measuring process is reproducible, the measuring instrument is reliable—this does not mean it is valid! It simply means the measurement instrument does not produce erratic and unpredictable results. It may be measuring a variable wrongly all the time but as long as it measures it consistently wrongly then it will be reliable! This may seem odd but what it basically means is that reliability is a necessary condition for validity but not a sufficient condition on its own. A very important aspect of reliability lies in the definitions of variables that are being measured. If we construct a variable such as 'sensitivity to prices' and ask respondents a series of questions in order to measure their price sensitivity, we need to be absolutely certain that we are measuring what we think we are measuring. Our very own definition of 'price sensitivity' may not be shared by all respondents or even with current theoretical understanding of the concept. For reliability in measurement, especially in survey research, we must have a clear and unambiguous definition of all the concepts and artificial constructs being used in the research design. If this is not the case, we will find it very difficult to make any kind of sensible and useful generalisations from our research findings.

There are two ways that reliability is usually assessed: first by checking the stability of measurement using the test–retest method (repeatability) and second by examining internal consistency applying the split-half method.

Test–retest Method

Assessing the repeatability of a measure is the first aspect of reliability. The test–retest method is a conservative method to estimate reliability. The idea behind it is that one should get the same score on test 1 as on test 2. There are three main components to this method.

1. Administering the measurement instrument for each subject at two separate times to test for stability. It is said to be reliable if the measure is stable over time. For example, suppose a researcher measures job satisfaction and finds that 60 per cent of the population is satisfied with their jobs. If the study is repeated a few weeks later under similar conditions, and if s/he finds the same result, it appears that the measure is reliable.
2. The high stability correlation or consistency between the two measures at time 1 and time 2 indicates a high degree of reliability.
3. Assume there is no change in the underlying conditions (or trait you are trying to measure) between test 1 and test 2. For example, at the individual level, we assume that a person does not change his or her attitude about the job.

When a measuring instrument produces unpredictable results from one to the next, the results are said to be unreliable because of error in measurement.

Split-half Method/Equivalent-form Method

The second dimension of reliability concerns the homogeneity of the measure. The technique of splitting halves is the most basic method for checking internal consistency when a measure contains a large number of items. In the split-half method one may calculate result from one-half of the scale items (e.g., odd-numbered items) and check them against the results from the other half of the items (e.g., even-numbered items).

However, in the equivalent-form method two alternative instruments are designed to be as equivalent as possible. Internal consistency estimates reliability by grouping questions in a questionnaire that measures the same concept. For example, you could write two sets of three questions that measure the same concept (say class participation) and after collecting the responses, run a correlation between those two groups of three questions to determine if your instrument is reliably measuring that concept. The closer the correlation coefficient is to 1, the higher the reliability estimate of the instrument. Both the split-half and equivalent-form methods measure homogeneity or internal consistency rather than stability over time.

14.3 VALIDITY

Validity is the strength of our conclusions, inferences or propositions. It involves the degree to which you are measuring what you are supposed to—more simply, the accuracy of your measurement. For instance, we are studying the effect of strict attendance policies on class participation. In this case, we saw that class participation did increase after the policy was established. Each type of validity would highlight a different aspect of the relationship between our treatment (strict attendance policy) and our observed outcome (increased class participation). There are four types of validity commonly examined in research methods.

1. Internal validity asks if there is a relationship between the program and the outcome we saw, is it a causal relationship? For example, did the attendance policy cause class participation to increase?
2. External validity refers to our ability to generalise the results of our study to other settings. In our example, could we generalise our results to other classrooms?
3. Construct validity is the hardest to understand. It asks if there is a relationship between how I operationalised my concepts in this study to the actual causal relationship I am trying to study. Or in the example, did our treatment (attendance policy) reflect the construction of attendance, and did our measured outcome—increased class participation—reflect the construct of participation? Overall, we are trying to generalise our conceptualised treatment and outcomes to broader constructs of the same concepts.
4. Conclusion validity asks if there is a relationship between the program and the observed outcome. Or, in the above example, is there a connection between the attendance policy and the increased participation?

It is believed that validity is more important than reliability because if an instrument does not accurately measure what it is supposed to, there is no reason to use it even if it measures consistently (reliably).

Threats to Internal Validity

Internal validity concerns the likelihood that changes in the dependent variable (the subject of the research) can only be attributed to manipulation of the independent variable and not to some other variable. When this is the case, a study is said to have high internal validity. If it is possible to provide an alternative explanation for the results of the study, the study has low internal validity. The following factors are important threats to the internal validity of a study.

History

History becomes a concern when external events affect the outcome of a study. For example, if a study were carried out on the risks associated with various financial institutions, a public crisis of confidence in the banking system during the period of the study would adversely affect the experimental outcomes. The longer a study lasts, the more likely it is that history will become a problem.

Maturation

Maturation refers to changes that can occur in the subjects of the study over a period of time. This includes ageing, fatigue and acquisition of skills or experience over time. For example, bank employees might become accustomed to a commission incentive after two weeks and stop trying to sell more.

Testing

Testing effects can be attributed to changes in subjects that arise from the influence of the testing process itself. For example, it is possible that a pre-test can sensitise or bias a subject's behaviour and result in an improved performance on the post-test. One way to overcome this threat is to use a different post-test to the pre-test.

Instrumentation

Instrumentation refers to inconsistency or unreliability in the measuring instruments or observation procedures during a study. For example, observers may be inconsistent in what they record during a study, or a post-test may be much more difficult than a pre-test.

Selection

Selection problems arise from one group in an experiment being different from another group. For example, one group might be brighter, more experienced or more receptive to change than another group. In other words, 'people factors' can cause a bias in the study. This threat is overcome by the random assignment of subjects to groups. It remains a problem where existing groups are used for the treatment and control groups.

Mortality

Mortality refers to the attrition of subjects from a study. The longer a study proceeds, the more likely it is that mortality will become a problem, especially if the subjects who drop out share a common characteristic with the entire group.

Threats to External Validity

External validity refers to the degree to which the findings of a research study can be generalised to other settings and situations. When conducting an experiment, a researcher hopes that the findings can be applied at a later time to other groups of people in other geographical locations. The following factors may threaten the external validity of the study.

Reactive Effects of Testing

The artificial effects of pre-testing may sensitise the subjects to the treatment. Without a pre-test, different outcomes may result from the experiment than would occur in practice. For example, if a study of attribute change was to begin with a pre-testing of attitudes, participants might become sensitised to the attributes in question and therefore show more attitude change as the result of the experimental treatment.

Reactive Effects of Selection

If the samples drawn for a study are dissimilar to the general population, it becomes difficult to generalise findings from the sample to the broader population. For example, the findings of a study involving only urban dwellers may not be applicable in rural settings. It is desirable to use samples that are representative of the broadest population possible. One widespread selection practice in much management research is to use university students as subjects. This practice poses a significant external validity threat to many studies if results are extended to the commercial world.

Reactive Effects of Experiment Setting

The arrangements for an experiment or the experience of participating in the experiment may limit the generalisability of the findings to other settings. Reactive effects can also occur when subjects know that they are participating in an experiment. This effect was demonstrated many years ago in the Hawthorne Plant of the Western Electric Company in the USA. A part of this study investigated the relationship between productivity and the brightness of lighting in the factory. As expected, productivity increased as illumination was increased. However, as brightness was decreased, productivity also rose. It was concluded that it was the attention the workers were receiving, rather than the lighting that was affecting production. This type of experimental participatory effect has become known as the *Hawthorne Effect*. The issue of validity can also be considered under several specific headings, which allow us to consider more carefully if a piece of research can be considered valid. These are presented in the following paragraphs.

Construct validity: Are we measuring what we think we are measuring? This seems a strange question but if you think about it we use many 'measures' of 'things' in the social sciences and in business—very often things that are not directly measurable. For example, in marketing we often refer to 'consumer satisfaction' using a range of 'indicators', but in fact these are artificial constructs—they are not real like a price, a metre or a velocity. In business we often refer to business 'confidence' of small firms, but one manager's confidence level cannot be directly compared to another's. Perhaps the most famous (or infamous) artificial construct ever applied in the social sciences was the Intelligence Quotient (the IQ test). This was used in schools all over the world for many years to 'select' those who could best benefit from a more academic type of education. In fact over 50 years of research showed unambiguously that the IQ test measured many things, but one thing it certainly did not measure was intelligence! Instead it measured the benefits of having books at home at an early age, of having educated parents, of having a caring and progressive environment at home—basically it measured the benefits to children of being born into a middle-class family. Therefore we need to be very careful about the constructs we use to measure anything in business or the social sciences and we need to be very confident we are really measuring what we think we are. Otherwise all of the traps we are aware of in logical reasoning and 'acceptable' evidence (see Chapter 2) will be waiting for us.

Convergent validity: Does it correlate with other measures we expect it to? The *a priori* test! This specifically refers to our theoretical and logical expectations for a measurement. A simple example is the relation between height and weight. We should expect *a priori*[1] taller people to be heavier than shorter people simply because they carry greater mass. If we find this is not the case then we have a very serious theoretical issue to consider and also a very serious measurement issue. In other words, we expect one measure (of a construct) to converge with another construct such as a rise in GDP should be positively correlated with a rise in living standards, an improvement in the health of an individual should converge with an increase in that person's sporting activities and so on.

Content validity: Does it measure/cover *all* the spectrum of the concept? For example our number system is based on a consecutive series (1, 2, 3, 4, 5, etc.), but if we only counted the odd numbers (1, 3, 5, etc.) then we are excluding half of the concept or construct. In a democratic political system if we only count those who actually vote at an election, then we ignore that portion of the voting population who have reasons not to vote. In using a 5-point semantic scale to measure opinions, if we exclude the midpoint (3) then we are preventing people from expressing the opinion that they are indifferent to the question even although they may well be indifferent! Just as in the case of construct validity we must ensure that any measure used to quantify a concept does in fact cover the full range that we can theoretically expect the concept to cover.

Predictive validity: Can the measure be correlated with other measures in forecasts? This is similar to the idea of convergent validity except that it explicitly refers to data that does not yet exist—in other words to forecasts. If construct A is convergent with construct B now then we should expect the future values of B to also be convergent with the future values of A. We can summarise the important role of validity in research in a simple diagram as shown in Chart 14.1.

We can only arrive at a conclusion that is both logically and materially true if we are confident that all of the steps in ensuring validity have been carefully understood. Even if this is the case then we face another problem in establishing both the reliability and validity of any research project—this is the problem of generalisability.

---

[1] *A Priori* is a Latin term meaning in the absence of evidence we will have a logical expectation that X is positively related to Y.

Chart 14.1 Steps to Achieving Validity

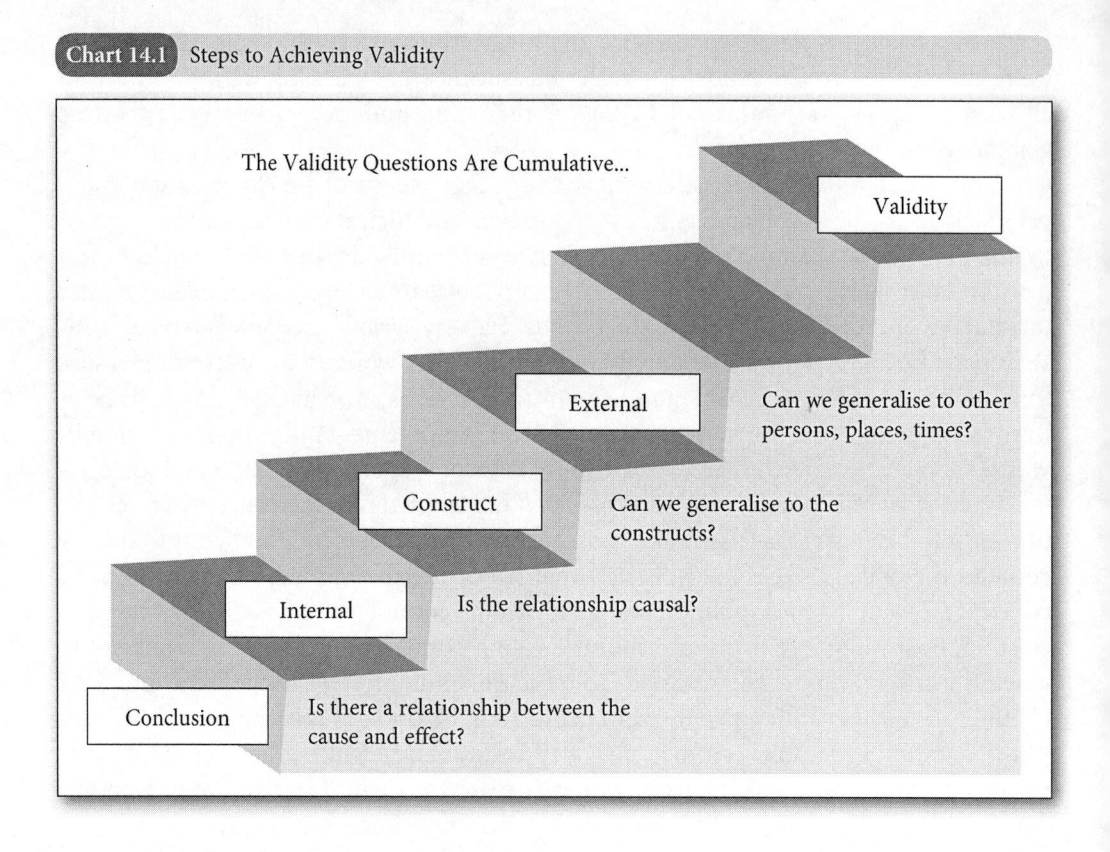

The Validity Questions Are Cumulative...

- Validity
- External — Can we generalise to other persons, places, times?
- Construct — Can we generalise to the constructs?
- Internal — Is the relationship causal?
- Conclusion — Is there a relationship between the cause and effect?

14.4 GENERALISABILITY

It is important that we are able to say something about a particular phenomenon, which is outside our own (narrow) research study. Why is this? Because unless we can make some generalisations then we are not really pushing knowledge forward and that is the whole point of research. Of course, in business research for example it is the case that a manager may not be interested in generalising research findings and only interested in 'solving' his or her particular research problem. That is fine in terms of immediate problem solving, but in terms of gaining a deeper understanding of the very nature of a particular problem it does not go very far. Most businesses will seek an 'off the peg' solution to a business problem because it is often cheaper than undertaking their own research. But such solutions only exist *if* the research which produced them was capable of generalising its findings. The sensitivity of research results to any small changes in any relevant factors also helps to establish the efficiency of the estimator for a certain proportion of data and this is very important for generalisation.

In essence, the ability of any research design to produce findings which are (mostly) applicable to other situations, organisations, countries and other people is dependent on the quality of the underlying theory, which allows us to interpret the 'world' in the context of a given research problem. Without theory no amount of empirical data will enable us to better understand the world we live in. That is the crux of generalisability—the ability to explain the same (or similar) phenomenon at all times and in all places without necessarily having to study it directly at all times and in all places. And of course the power of generalisability is greatly increased where both reliability and validity have been assured.

14.5 EXERCISES

Exercise 1

Provide an example of a cross-sectional study that you are familiar with. How do you think the researchers attempted to make the study representative? Give an example of a longitudinal study that you would wish to carry out. What particular problems might you encounter?

Exercise 2

Scenario: Two studies were carried out to establish if there is a link between smoking and certain types of cancer. One study was conducted in the 1950s and compared mortality rates for ex-servicemen with the general population. The second study was carried out in the early 1990s and compared the health of girls and young women aged 16–23 years who smoked, with a control group of girls of the same age who did not smoke. Neither study found evidence of a significant relationship between smoking and the certain types of cancer investigated.
   What validity, reliability and generalisability issues do the two studies throw up?
   In what way (if any) can the findings from the studies be compared?

14.6 WEBSITES

http://faculty.ncwc.edu/TOConnor/308/308lect04.htm
http://writing.colostate.edu/guides/research/relval/
http://www.georgetown.edu/departments/psychology/researchmethods/researchanddesign/validityandreliability.htm

CHAPTER 15

Conducting Your Research

15.1 INTRODUCTION

In this chapter suggestions on 'how to select a research topic' are given. After this, an outline of what supervision practice normally involves is presented. Then commentary on various practical issues on how to undertake your research is presented, focusing on project planning. This is followed by an example of a very important part of doing research—the research project proposal. Many students begin with a vague idea and do not fully appreciate the importance of proper research planning. The example shows why such planning is so important. The next chapter concludes with some advice on writing your dissertation, emphasising the importance of good structure.

15.2 SELECTING THE TOPIC

The following are a set of guidelines you might wish to follow in the process of deciding which research topic you want to focus on for your dissertation.

Focusing on a Research Topic

A very common problem with the selection of a research topic is that the 'idea' can often be very vague at the outset and is not sufficiently developed during the early stages of the research. If the 'idea' is not sufficiently developed and focused, the outcome can often be a mix of the following or even all of the following:

- Lack of direction in the Introduction to the dissertation
- Poor matching of initial objectives and outcomes
- Irrelevant material directed at issues that have little relevance to the research topic

To minimise the chances of this happening a useful strategy is to start at a broad level of analysis and work down; in other words, to *scope* the research 'idea' in such a way that it is feasible to actually carry out the research. One approach to scoping your research ideas is as follows:

Choose a research **'AREA'** defined in terms of a broad field of study such as demography, economic growth, international marketing, sociology, urban planning and so on. For example, consider a 'vague' idea for a dissertation about the relationship between population and economic growth.

Identify a particular **'FIELD'** within the broad area—for example, 'human capital accumulation within the population' would be a field within the general area of population and economic growth. Next, we need to consider what specific aspect of this 'field' we wish to investigate.

The specific **'ASPECT'** could be the relationship between years of schooling and GDP growth across a number of countries and/or across a number of years. In Figure 15.1 the student has successfully narrowed down the field to a specific aspect that is manageable and do-able.

Or it may be that a particular technique or strategic tool has interested you in your studies. For example a tool such as The Ansoff Matrix can feature strongly in a project that examines the different strategies a company can adopt if their priority is growth. The product–market matrix proposed by Ansoff provides a useful framework for considering strategic direction and marketing strategy for the company. Whatever your subject discipline, you need to identify the area first, then the field, followed by a specific aspect of that field. This process will enable you to better focus your research purpose, research aim and specific research questions and/or hypotheses.

Does It Interest You?

You are going to be 'living' with your choice of research topic for quite a while; therefore, it is very important that you choose something which genuinely does interest you. If you are not particularly motivated by the subject area you are likely to do the minimum required to complete the research. Ideally the research topic should be one which makes and keeps you curious and one which you feel will also interest the reader. Make a note of ideas and interesting concepts as they occur to you. Try to collect ideas well ahead of

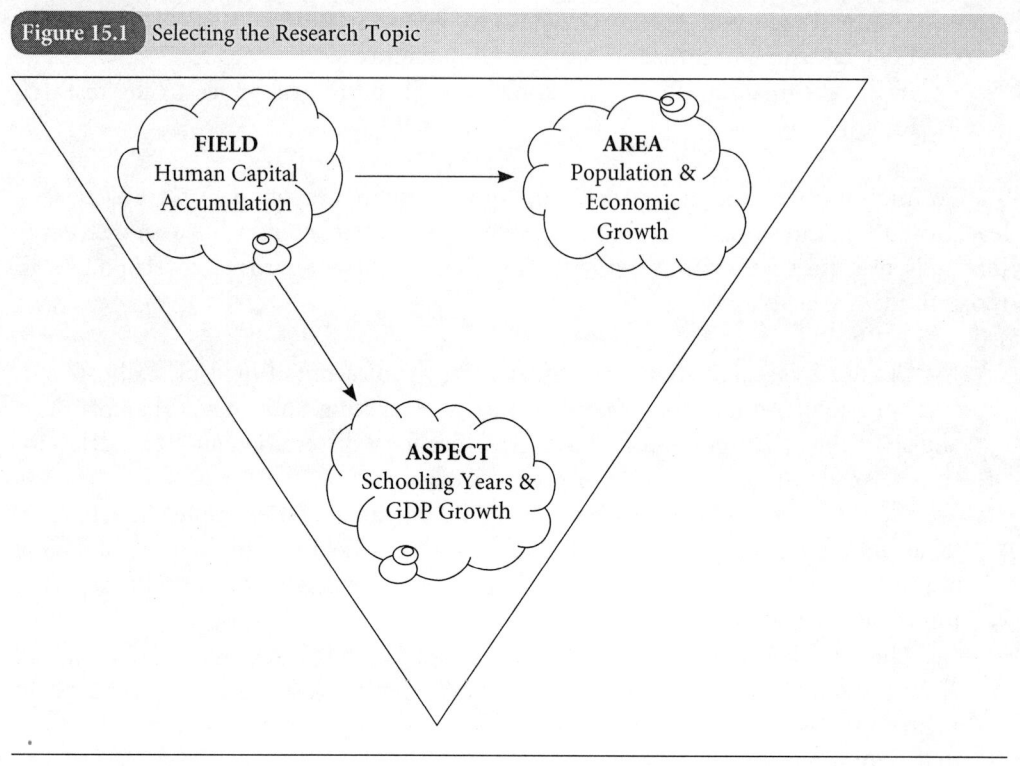

Figure 15.1 Selecting the Research Topic

- **FIELD**: Human Capital Accumulation
- **AREA**: Population & Economic Growth
- **ASPECT**: Schooling Years & GDP Growth

SOURCE: Authors' own.

the deadline for your research proposal. Social issues in marketing are of interest to many students and marketing practitioners.

How Much Do You Know about It Already?

Clearly you will have less work to do if you choose a topic you've encountered and worked on before. You will be more aware of the issues that arise if you've covered the basic groundwork already and analysis and evaluation should be more straightforward. Any topics that you have already researched (perhaps for an assignment) are worth considering as potential research project subjects, but you must be sure that they have the depth to be developed. For many students your work experience can be a very useful basis for topic choice and identification since it could be related to your own organisation. Be aware though that besides being relevant and drawn from your own experience it must also be academically rigorous. In terms of your professional development you might consider a topic which helps you to develop in a desirable direction. For example, you could ask yourself, "what sort of topic will help me with my career plans?"

How Difficult Is a Dissertation Likely to Be?

'Difficulty' in this sense refers to the following:

The level of qualification: At the undergraduate level to show understanding of the topic and competency in applying ideas from your programme are the key expectations. Collecting and analysing the data are often at the level of illustrating what can be done. The analysis should be competent and you need to make conclusions which can be supportive from your research. Thus, the conclusions tend not to be firm and limited. Caveats often need to be added that the data collected is limited. At Masters level your dissertation should reflect a clear understanding of research method, the application of relevant analytical techniques and a clearly focused and well-argued analysis of the specific research questions you set out in your research proposal. Your choice of topic should allow you to meet this standard.

Complexity of the subject matter: Some topics are very difficult to handle and should be cautioned against if there is relatively little material available in libraries or other sources including your lectures or tutorials. Similarly if your 'idea' or refined topic is rarely referred to in textbooks and journals this is usually an indication that you will struggle to find relevant research material on the topic.

Availability of expertise: As well as library, textbook and journal material, 'in company'-based research can benefit from the knowledge and experience of employees who are familiar with the issues which your topic addresses. If there is such a person this might influence your choice of topic but remember that the research itself and the dissertation are your responsibility—do not over-depend on other people. The author of the following dissertation was employed in an advertising agency and was making use of his own and others' knowledge of the industry: "The impact of the relationship between account planners and creative personnel in Advertising agencies".

Another student was employed by one of the leading life assurance companies world-wide and had been interested in her chosen topic for some time. "Are consumers making an informed choice when purchasing life assurance in today's financial services market?"

Ease of data access: Research at any level normally requires you to present empirical material. This can be original or derived from secondary sources. Is your topic one for which data is easily obtained or found? If the data is too difficult to generate, too expensive to collect or takes too long to acquire you might do best to rethink your choice of topic. You may also need to rethink if there is no significant body of literature to investigate. A survey-based topic will require you to be careful in drawing up a sample frame from which you decide on the particular sample of respondents to approach. Your choice of topic will need you to make an early decision on whether a sampling frame already exists or needs to be created by you.

Time required for completion: This can be influenced by the complexity of the topic and a timetable imposed on you by events often outside your control. Care has to be taken in estimating both timing and feasibility. If you are likely to have to travel to interview respondents or spend a lot of time accessing data then planning needs to be done at an early stage. This is also a critical part of the 'scoping' you should undertake at the outset.

What do you wish to do when you graduate? On finishing your present programme what you plan to do is an important concept to think about. Undertaking the dissertation allows you to find out about an area of interest and allows you to appear knowledgeable at any interview. The research gives you a taste of what more research would be like such as undertaking a Master's degree or even a doctorate. A dissertation represents a substantial body of your own work and can be shown to potential employers and mentioned on your curriculum vitae so it can aid future career plans.

In summary a number of key elements should be kept in mind when deciding on your choice of research topic for your dissertation:

- Your interest and curiosity in the subject
- The level of prior knowledge you have
- Suitability for the level of your degree
- The availability of expertise
- The ease of data access and data availability
- The time required for completion
- Your future plans

In undertaking your dissertation, the single most important resources for you is your supervisor and there now follows some guidelines for the supervision process.

15.3 GUIDE TO SUPERVISION

Using Your Supervisor's Time

In most educational programmes that involve a dissertation each student is allocated a supervisor who in turn is allocated a certain amount of time for the supervision and you should try to utilise this time in an efficient way.

There are a number of areas on which you can call on your supervisor for help, for example:

- Aid you in *refining* your choice of research topic
- Advise on relevant academic literature
- Discuss and advise on the design of the research and the proposed research method

- Advise on the relevance and practicality of the proposal
- Agree the timetable for completion
- Discuss the research findings/problems to date
- Read and discuss draft chapters and one complete draft

Try to make sure you contact your supervisor several times before you submit the final draft of your dissertation. If you have worked on and written up a dissertation completely unaided, there is a significant danger that you will have made errors or omissions, which will require a considerable amount of work to rectify.

Whilst your supervisor will attempt to answer queries large and small, he or she will not expect to be answering your emails several times a week for lots of weeks in succession. It is likely that there will be periods when you will wish to ask questions frequently, but these should be counterbalanced by periods when you are working on your own by yourself with little contact. Your supervisor will use his/her discretion in deciding what he or she can usefully help you with and where you should be working on your own initiative.

Your supervisor will be happy to review a draft of each chapter if you wish to submit them as you go along, but, having commented once, would not expect to see a redraft of that chapter unless you had to substantially rewrite it. He or she will of course review a written draft of your dissertation, which you may submit by the due date and will annotate it and point out any major problems. However, this is not a pre-marking and your supervisor will not tell you at that stage what mark you can expect to get.

Bear in mind, also, that the supervisor's advice is exactly that. You should not expect him or her to provide a definitive list of references for you to seek out or to write a questionnaire for you. Part of the value of doing a dissertation is that you do these things yourself. The dissertation is ultimately your responsibility!

15.4 UNDERTAKING YOUR RESEARCH

Once you embark on your dissertation time quickly passes, so it is essential to plan what you are going to do. An activity on (or before) starting should be to plan your dissertation and develop time-based milestones for when key tasks should be completed. To form a detailed time plan, and stick to it is of great help and should be laid out with help from your supervisor. To form a plan, first identify all the activities—the more detailed this can be, the better. Then attach time duration to these activities. For example consider a study into how the culture of restaurants affects performance and it is to be undertaken in 140 days. Perhaps three case studies will be undertaken. A table such as that displayed in Table 15.1 might be formed. In this each activity is given an alphabetical code.

Table 15.1 Activities and Durations of Culture and Performance Project

Activity	Code	Duration
Initial meeting	A	1 day
Obtaining and reviewing literature	B	30 days (although goes on for the duration of the study)
Determining aims and research questions	C	2 days
Meet supervisor	D	1 day
Contacting companies to allow study	E	5 days
Complete literature review	F	5 days
Case study 1	G	15 days
Write up case	H	5 days
Complete research methodology chapter	I	10 days
Case study 2	J	15 days
Write up case	K	5 days
Case study 3	L	10 days
Write up case	M	5 days
Meet supervisor	N	1 day
Compare and contrast cases and relate to literature	O	6 days
Write chapter on comparing cases	P	4 days
Meet supervisor to discuss conclusions	Q	1 day
Write conclusions and discussion	R	3 days
Write introduction	S	2 days
Present draft	T	0
Write abstract	U	1 day
Improve diagrams and ensure references are in order	V	3 days
Meet supervisor	X	1 day (but allow at least 14 days for feedback)
Final amendments	Y	5 days
Produce contents and acknowledgements	Z	2 days
Submit dissertation		

SOURCE: Authors' own.

Note that although 30 days are allowed for the literature review it actually continues for the duration of your study. Five meetings with a supervisor are indicated, these can be considered milestones and in reality more may be had but rarely fewer, if good practice is adopted. Aim to submit the dissertation about two weeks before the final submission data to allow a bit of spare time in case final word processing problems, etc., arise. To organise this, a project plan such as in Figure 15.2 helps.

On the chart, one can go through indicating the earliest start time of each activity. This is called a 'forward pass'. This is illustrated in Figure 15.3 and the times are shown in the boxes in the figure.

This is then done in reverse, starting with the latest finish time of the preceding activity. These times are placed in the empty box in a procedure called 'backward pass'. If approximated two weeks are needed for contingencies then the latest finish time is 126 days. The backward pass is shown in Figure 15.4.

When boxes have the same duration on either side then the activities linking them are said to be critical. Any delay in a critical activity will lengthen the project. If you do your plan and you find you do not estimate completion on time, examine the critical activities to determine which can be shortened, perhaps by being less ambitious. Activities not on the critical path have 'floats', which is the latest end time minus the

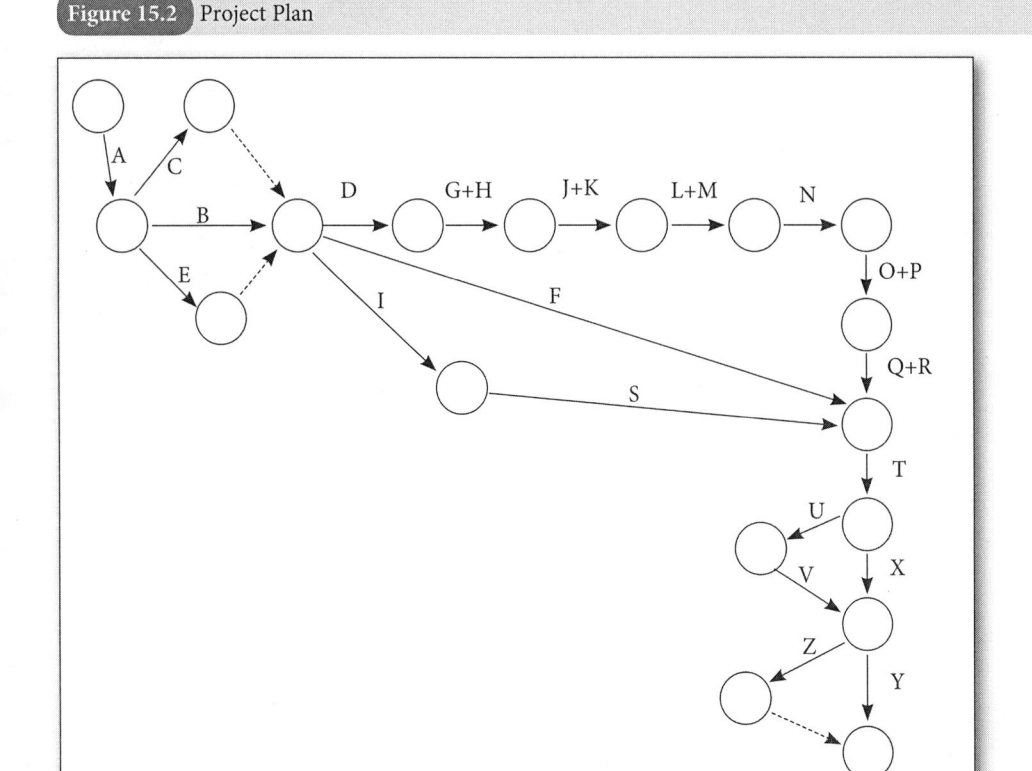

Figure 15.2 Project Plan

SOURCE: Authors' own.

Figure 15.3 Project Plan with Forward Pass

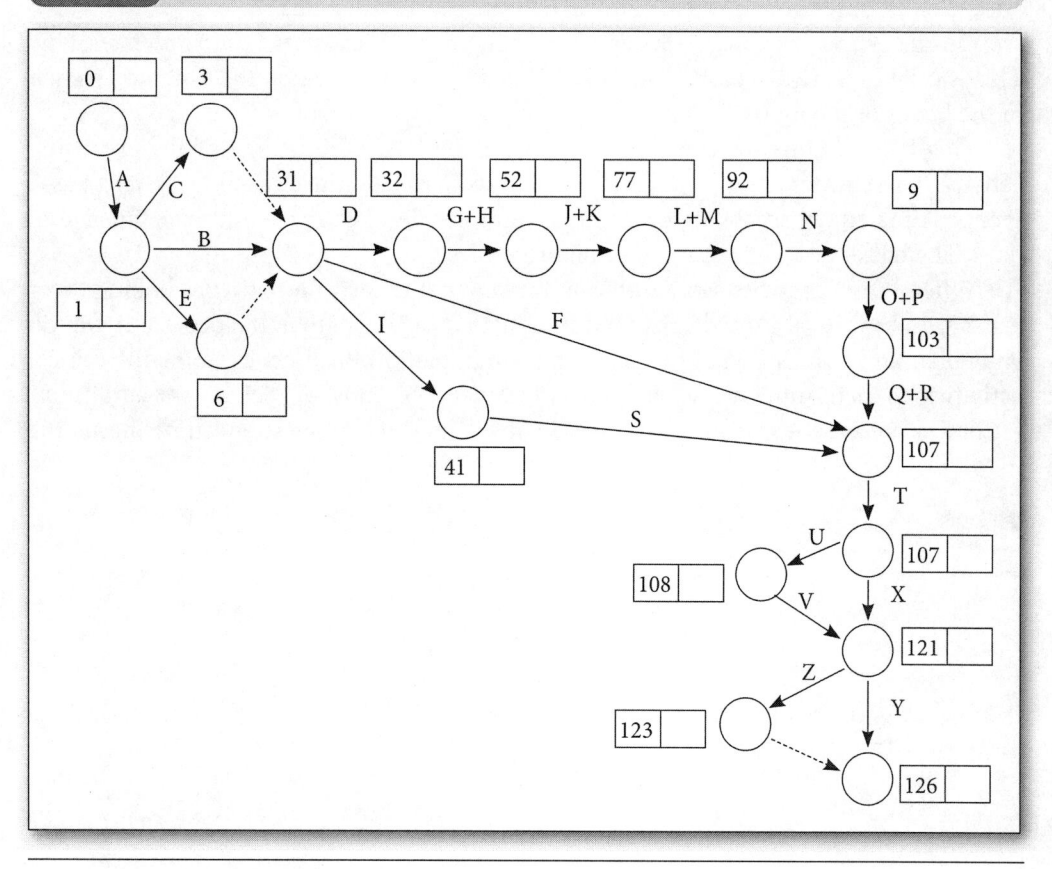

SOURCE: Authors' own.

earliest start time and minus the duration of that activity. The start and time can be varied to the magnitude of the float.

After doing this a Gantt chart can be displayed, which is a time chart marking the critical path and showing the milestones, which in this project will be the meetings with the supervisor (for more information see Burke [2003] and http://www.mindtools.com/pages/article/newPPM_03.htm).

In Figure 15.5 the Gantt chart of this project is presented.

The black boxes represent the duration of the critical path and the grey boxes show activities with floats.

If you need to access a company for interviews or a case study or need some other support, ensure that you get their co-operation as soon as possible and within a few weeks of starting. Try to obtain a letter stating that co-operation will be given. This helps if your contact leaves the company and is good practice. These letters can be placed in your dissertation.

Figure 15.4 Project Plan with Time Determined

SOURCE: Authors' own.

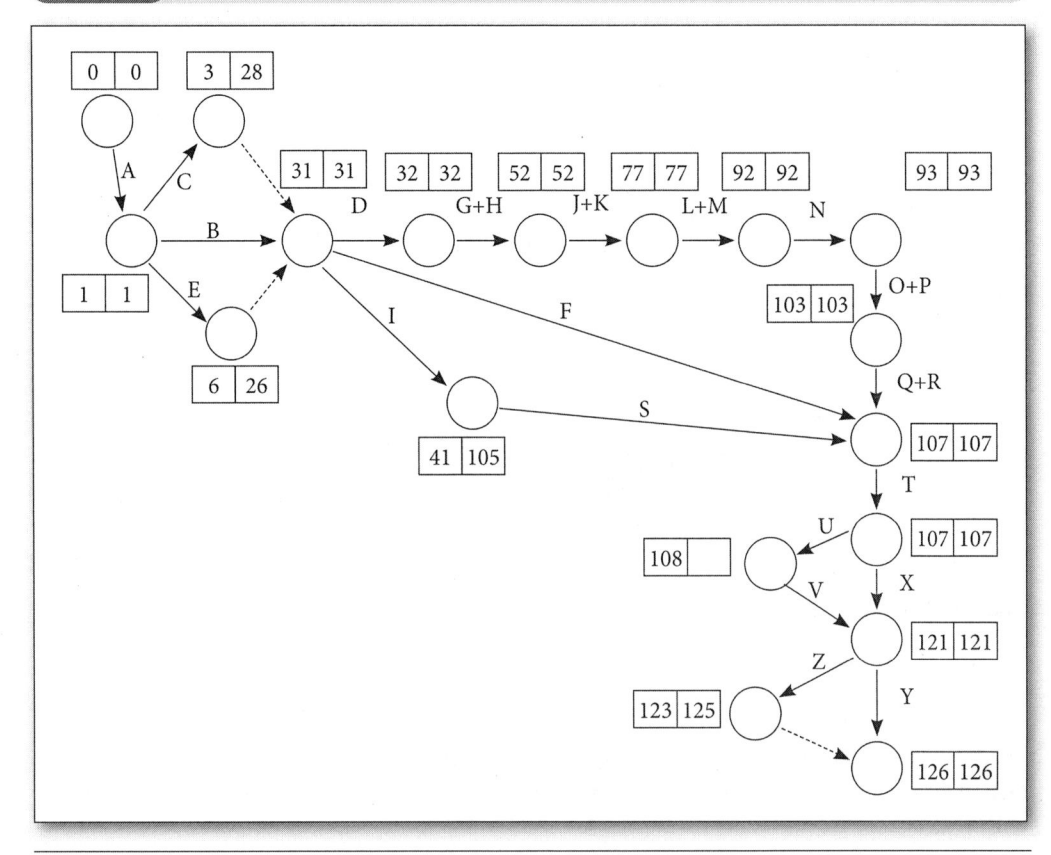

Another good practice is to keep a diary or a notebook to document your progress and jot down any observations or additional information you find. Similarly, it is good practice to write up minutes of meeting with your supervisor and to prepare action plans from each meeting. These you may like to be agreed by your supervisor. Supervision meetings will always be more useful if you prepare for them and if possible send material in advance.

15.5 RESEARCH PROPOSAL EXAMPLE

The following example is not perfect; however, it does include all of the elements that a supervisor would expect to see in a research proposal for any level of degree (under and postgraduate). As the reader you need not be concerned with the specific subject matter

Figure 15.5 Gantt Chart

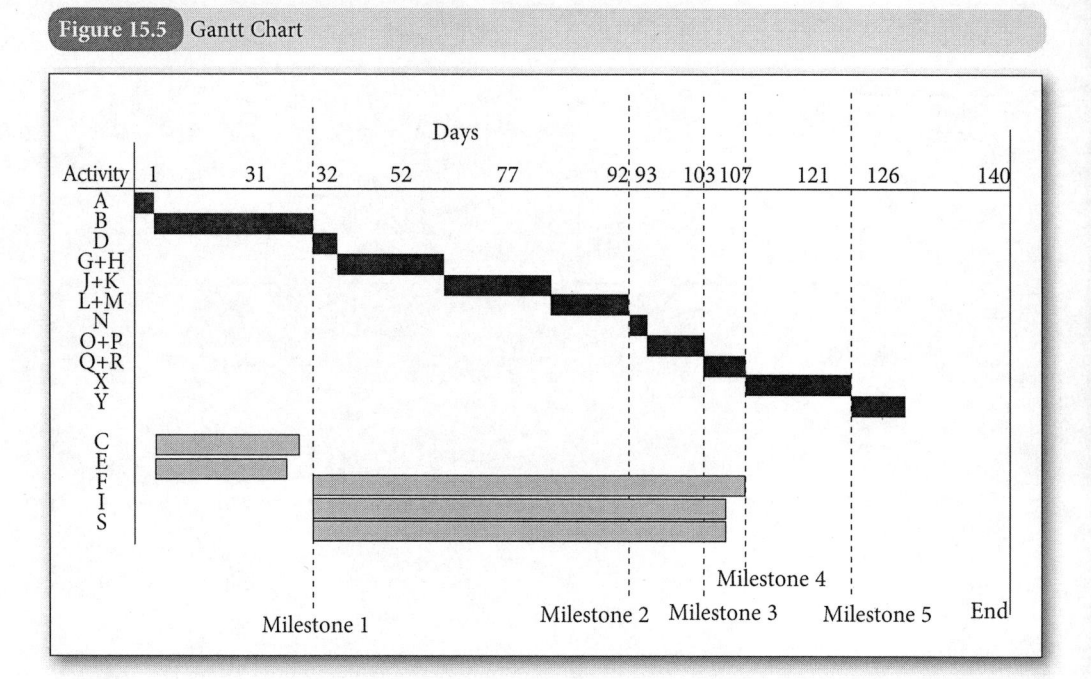

```
                              Days
Activity  1      31  32    52        77        92 93  103 107    121    126        140
A         ■■■■■■■■■■
B                    ■■■■■■
D                           ■■■■■■■■■■
G+H                                   ■■■■■■■■■■
J+K                                             ■■
L+M                                                   ■■■■
N                                                         ■■■■■■
O+P                                                              ■■■■■
Q+R                                                                     ■■■■■■■■■
X
Y                                                                                  ■■

C         ▭▭▭▭▭▭▭▭▭▭
E          ▭▭▭▭▭▭▭▭▭
F                          ▭▭▭▭▭▭▭▭▭▭▭▭▭▭▭▭▭▭▭▭▭▭▭▭▭▭▭▭
I                          ▭▭▭▭▭▭▭▭▭▭▭▭▭▭▭▭▭▭▭▭▭▭▭▭▭▭▭▭
S

                                                     Milestone 4
         Milestone 1         Milestone 2  Milestone 3           Milestone 5   End
```

SOURCE: Authors' own.

itself, instead you should try to understand and appreciate the structure of the proposal, how it has been written and put together and the various elements that are sub-headed within it.

THE BRITISH UNIVERSITY IN EGYPT

Dissertation Proposal

Democracy and Economic Growth

BSc Economics, Degree Year 2

Abstract: In an attempt to clarify the relationship between democracy and economic growth, I propose this document to have the approval to start my dissertation topic. The proposal includes the main aims of the dissertation topic, a brief literature review, data collection method and data analysis method. The proposal will show what research methods I will be using to undertake my dissertation research. There is a section that discusses the limitations of the research topic and an estimated budget and costs for the dissertation is also included.

Name:	Nevine Essam
ID:	
Area of Economics	Economic Growth, Politics and Development
Main Research Approach	Quantitative and Qualitative approaches
Main Data Collection Method	For quantitative data: Secondary data collection method from databases, government agencies and reports. For qualitative data: Secondary data collection method from books, journal articles, official reports and government papers.
Country, Industry, Product or Service Sector you will focus on	A sample of countries will be selected and focusing on their political approaches (democracy, authoritarian, etc.) and its effect on economic growth. Political systems and the economy.

CONTENTS:

I. Introduction
II. Aims and specific objectives
III. A brief literature review
IV. Research methods
V. The research process
VI. Limitations of the research
VII. Estimated Budget
VIII. References

I. Introduction:

In this dissertation I will investigate and analyze the relationship between democracy and economic growth. This topic had been discussed before by many economists and writers; however, there is no final precise result that shows whether this relation is positive or negative. Although many writers and economists concluded that democracy is a major prerequisite for economic development and growth, others have reached contradicting conclusions where sometimes an authoritarian regime has the best economic outcome. Through analyzing and comparing data of different countries with different types of regimes (democracies or dictatorships), I hope to be able to reduce the ambiguity of this relationship. This topic is of major importance because it addresses one of the major questions in the science of political economy.

As Abeyasinghe, R (2004) stated, this question is of major importance for developing countries because many international institutions that aid developing countries like the IMF and the World Bank set political liberalization and democracy as a precondition for obtaining their aid. If analyzed precisely and accurately, this topic will be able to give signals to the best political system a country can implement to improve its economic performance. If it is successfully shown that an authoritarian or dictatorship regime is better for the economy and the welfare of world's citizens, the way people perceive such regimes may change. Psychologically, citizens may be able to accept the fact that an authoritarian regime may be best (for a while) as the most efficient choice for them. On the other hand, if it is successfully shown that an authoritarian regime hinders economic performance, this will assure us that such regimes should be reformed to democratic ones. Of course the above results are not easy to achieve because there are millions of other factors that lead to them. In fact, it is not like this dissertation will be able to successfully do all of the above, because realistically, it will not. However, this dissertation will try to make a small contribution to help other researchers and students to understand what the topic is all about. The analysis and the results of this dissertation will help anyone who is interested in this topic and maybe someday it will be a part in the reason for someone to find a convincing answer.

II. Aims and Specific Objective:

'Does democracy hinder or help economic performance?'
The answer to the above question is the main purpose that this dissertation is trying to achieve through historical analysis and through gathering and comparing data of different countries in the world. Sharma, S. D. (2007) argues that there is a 'belief' that the more democratic a country is the better its economy is performing. On the other hand, some people believe that this is just an illusion. Doucouliagos, H. & Ulubasoglu, M.A. (2008) believed that although democracy benefits the economy, dictatorship can be more beneficial. This is what the dissertation will be trying to analyze. In the dissertation, I will try to reach a certain answer to some research questions. Does democracy benefit the economy? Or, does it harm the economy? Is an authoritarian regime more beneficial to the economy than a democratic one? Does the hypothesis differ from one country to another? Do developing countries really need an authoritarian regime to improve their economy, or not? Are the economies of these countries not ready for democracy yet? Can developing countries afford the economic costs of democracy? The same questions will be asked in the case of developed countries. At the end, we will be able to know if democracy is good or bad for an economy.

In the dissertation I will try to answer all of the above questions through giving real examples, focusing on both developing and developed countries over a about a 50 year time period. I will also give examples of developed democratic countries and developed authoritarian countries, and compare their economic performance. If followed accurately, the above process will enable me to reach a clear conclusion of what appears to be best for each—the developing and the developed countries.

III. A Brief Literature Review:

In this section, I will try to collect the point of views of different writers about the effect of the type of political regime, democracy or dictatorship, on economic growth. Doucouliagos, H. & Ulubasoglu, M.A. (2008) showed a comparison between the pro and anti-view of democracy. They stated that the pro-democratic views show that under a democratic regime, citizens have more incentive to work and invest under the protection of property rights and the market will be able to efficiently allocate the resources. A democracy limits the degree of state intervention in the market place so that it encourages a stable and sustainable economic growth in the long run. Sharma, S. D. (2007) argues that a democracy offers better long term protection of property rights and freedoms. Heo U. & Tan A.C. (2001) show that the protection of property rights under a democratic regime definitely encourages production and the exchange of economic goods. In fact, they stated that "nations that protect property rights tend to grow faster than those that do not and that democratic societies tend to protect property rights more than other types of government".

Mahmoud K., Azid T., & Siddiqui M.M. (2010) argue that democracy enhances the political life through free elections that do not differentiate between social groups, this in turn leads to a good degree of political and civil rights that enhances competition and so improves the economic state of the country. An efficient democratic regime that responds quickly to the demands of the public is needed for economic development. However, democracy also has its negatives that greatly affect the economy of any country, Doucouliagos, H. & Ulubasoglu, M.A. (2008) explain that often democracies have weak and fragile political institutions, which means they 'lend themselves to popular demands at the expense of profitable investments'. They argue that some of the negatives of a democratic regime, where a country is prone to social, ethnic or class conflicts, can hinder economic growth. Due to the frequent changes in the governments through elections in democratic regimes, the economic policies set by one government are usually ended when the new government is elected, which leads to immature and uncertain economic policies. Mahmoud K. et al (2010) argue that these conflicts cannot be efficiently suppressed by democracies, which shows that in order for democracy to function properly, a certain level of development must be reached as a prerequisite.

In Doucouliagos, H. & Ulubasoglu, M.A. (2008) point of view; "Political democracy is a luxury good that cannot be afforded by developing countries". On the other hand, they suggest that, though strict and tough, authoritarian regimes are able to suppress social conflicts and are able to take strict measures needed for rapid economic growth. In other words, an authoritarian rule will tend to improve the public welfare because it is probably the only way that effective and rapid solutions can be introduced to improve economic performance. Sharma, S. D. (2007) however strongly argues that *stable* democracies are the type of regimes that make a good economic performance across history. In Sharma's point of view, a durable democracy is strongly related to improving the economic performance.

Xinsheng, Q. (2007) is another author who sees an authoritarian regime as pushing economic development forward. His study shows that when the power and authority is centralized to a single body, the decisions of this single ruling body will be implemented smoothly and easily; which in turn increases efficiency in the production process.

But wherever the authority and the power are divided across several bodies and based on a democratic system, the decision making process is diversified and so implementing these decisions is not efficient enough to improve the productivity and efficiency of the economy. This suggests that the claim that during transitional periods, the economic growth of a country needs an efficient system of decision making which is usually present in an authoritarian regime. In fact this relation has been well demonstrated in history where the economies of Indonesia, Chile and Philippines had all witnessed rapid growth under the dictatorships of Suharto, Pinochet and Marco respectively. Although these countries' economies continued to grow after they became democracies, Xinsheng, Q. (2007) argues that the economic growth in these countries was not as fast as it was under the authoritarian rule.

Abeyasinghe, R. (2004) argues that the availability of democracy in developing countries can lead to the formation of economic policies that hinders the growth rate. For instance, democracy can prevent the implementation of policies that encourages trade liberalization that could have benefited the economy. Abeyasinghe, R. (2004) gave South Korea as an example, while the government was moving towards more political and economic freedom by the end of the 1980s, the farmers revolted against the free market policies like the liberalization of imports and so prevented the government from implementing policies that could have benefited the economic growth of South Korea. On the other hand, Artige, L. (2004) shows that a dictatorial rule can either form an economic miracle or a disaster. Although many statistics show that the world's poorest countries are dictatorships, it is obvious that the economic growth rates under this type of regime are either very high or very low. Artige, L. (2004) gives China as an example of an economically successful authoritarian regime where the Chinese GDP per capita had been doubling every 9 years since the application of its economic reforms, but no significant democratic reforms.

Kurzman C., Verum R. & Burkhart R. (2002) argue that poor countries' way to substantial economic growth is a "developmental dictatorship" regime, where the citizens must be obliged to work, sacrifice and obey the orders that a dictator sets in the favour of the country's economy. In analysing the effects of this type of political regime on the economy, they compared three elements of economic growth; investment, state expenditure and social unrest. For instance, investments flourish in democracies because the climate of liberty helps investment to grow through the free flow of information and the protection of property rights.

On the other hand, investments can suffer in democracies because people are not obliged to change their consumption patterns in order to induce more saving and investments, in fact this change in the pattern of savings and investments require a long period of time. Through state expenditure they also argue that social unrest can be effectively minimised as long as investment can be seen to benefit the public. The latter is only a brief review of the literature in this field and the dissertation will significantly expand this.

IV. Research Methods:

The dissertation will be an applied longitudinal research as I will be investigating an area of research that aims at providing answers to whether democracy improves or hinders the economy. The longitudinal approach will be used particularly in analyzing the data. The dissertation will focus on growth rate indicators (GDP, net income, investment, government expenditure, income per capita, savings, investment and consumption) of several countries (such as China, India and others) over several years. It will also include measurements of democracy using the Polity Index (PI) or Freedom House Index (FHI) over the same period of time that the growth indicators will be measured in. The longitudinal approach will include the panel type of data as I will collect data for the same countries over a certain period of time. The dissertation will mainly try to test the hypothesis that democracy improves economic growth, which will make it more of a theoretical-deductive study than an empirical (inductive) one. Through getting data and information from a number of countries, the dissertation will try to draw general conclusions from these observations in order to demonstrate the validity of the theory and the significance of certain hypotheses.

V. Doing the Research:

Concerning the process of collecting data, I will use the secondary data approach i.e. government agencies, media, public libraries and the web. Data will also be collected from Journal articles, books, magazines and news and government reports. Specifically, quantitative data will be collected from government agencies and official databases, while the qualitative data will be collected from journal articles, books, news and libraries.

The quantitative data will mainly be data that measures economic growth of several countries in terms of GDP, net income, investment, government expenditure, income per capita, savings, investment and consumption for several years. These data can be easily obtained through various online databases provided by the World Bank, International Monetary Fund (IMF), the United Nations (UN) and the Human Development Reports (HDR). For measuring the democracy of a country, I will use the Polity IV index 1800-2010, in addition to the availability of the Democracy Index that was developed by the Economist Intelligence Unit (EIU). However, the EIU's Democracy Index was published for the years 2007–2010, which is considered to be a short period of time for the efficiency of data. Below are the websites of some of the data bases that will be used in collecting the economic data:

- http://hdr.undp.org/en/
- http://www.worldbank.org/
- http://www.un.org/en/
- http://www.imf.org/external/index.htm

The websites for collecting democracy measures will be:

- http://www.systemicpeace.org/polity/polity4.htm   http://www.systemicpeace.org/inscr/inscr.htm
- http://www.eiu.com/
- http://www.democracybarometer.org/start_en.html

Concerning the qualitative data, the above databases also provide many useful reports for getting a theoretical insight about the democracy–growth relation and its history. For literature and theory I will be using books, academic journal articles and official reports. Here are some of the websites that will be used:

- http://books.google.com/
- http://www.wri.org/
- http://www.oxfordbusinessgroup.com/
- http://scholar.google.com/
- http://lib.bue.edu.eg/wiki/index.php/Online_Databases
- http://www.economist.com/

In addition to these sites, I will also try to visit libraries in Egypt and Official Organizations that might provide me with data. The data analysis method will be a statistical and econometric one (time series), where tests of the hypothesis can be undertaken. In addition, graphical representation of data will be used to demonstrate what are the more obvious relations in the dataset.

VI. Limitations of the Research:

The dissertation data and its analysis are expected to be reliable because the data will be collected in a way that it includes a large number of countries through a long time period. The data analysis tools will allow us to have a clear view of the results, and as I have a large set of data that covers a long period of time, the measurement instrument in the dissertation gives a large chance to be reliable. When it comes to the validity of the dissertation, it will be assessed as follows. The internal validity of the topic is actually aiming at solving a question of causation, i.e. Does democracy cause or hinder economic growth?'

The external validity of this topic is strongly related to the reliability of the data; as said before, the available data are reliable as it measures democracy and economic growth of many countries through different years, which helps in generalizing the conclusions on other countries other than those included in the dataset. However, the external validity of my dissertation may be at a threat if the results of the tests are insignificant, which will prevent the generalizing of the conclusions to different countries. In addition to this, as economic growth is affected by many factors in addition to the type of the political regime, the conclusion may not be generalizable.

However, because of the availability of a large set of data I do remain confident that the generalizability of my conclusions will be stronger than if I only focused on a few countries. The topic is focused on the issue of explaining the phenomenon of democracy and economic growth in many countries over many years.

VII. Budget:

This section is for estimating some of the expenses that will be incurred doing the dissertation process.

- Travel expenses: I might need to travel to Alexandria to visit the Bibliotheca Alexandrina in order to get some sources for my topic. This is estimated to cost L.E. 600.
- Stationary Expenses: Photocopying papers and articles will cost approximately L.E. 200.
- Resources Expenses: The estimated cost for purchasing resources (i.e. books, journals) is L.E. 400.

VIII. References:

Artige, L. (2004). *On dictatorship, economic development and stability.* (Universitat Autuonoma de Barcelona).
Abeyasinghe, R. (2004). *Democracy, political stability, and developing country growth: Theory and evidence.* (Honors Projects, Illinois Wesleyan University). Retrieved from: http://digitalcommons.iwu.edu/cgi/viewcontent.cgi?article=1000&context=econ_honproj
Doucouliagos, H., & Ulubasoglu, M. A. (2008). Democracy and economic growth: A meta-analysis. *American journal of political science.* Volume 52, issue no. (1), 61-83.
Heo, U., & Tan, A. C. (2001). Democracy and economic growth: A casual analysis. *Comparative politics.* Volume 33, issue no. (4), 463-473.

> Mahmoud, K., Azid, T., & Siddiqui, M. M. (2010). Democracy and economic growth in Pakistan. *Research journal of international studies*, issue no. (15), 77-86.
> Kurzman, C., Verum, R., & Burkhart, R. (2002). Democracy's effect on economic growth: A pooled time series analysis 1951-1980. *Studies in comparative international development.* Volume 37, issue no. (1), P.3-33.
> Sharma, S. D. (2007). Democracy, good governance and economic development. *Taiwan Journal of Democracy.* Volume 3, issue no. (1), 29-62.
> Xinsheng, Q. (2007). Two essays on the market economy and democracy. *HRI.*

The above research proposal is well written, is logically structured and contains a structured abstract at the beginning, allowing you to be able to summarise very clearly the area, the field and the specific aspect of the research that is to be investigated. If you can produce a research proposal like the one above, then you have already gone a long way towards doing a successful dissertation. This particular proposal received a *first class* grading. In the final chapter (next), we present some ideas for you on how to write and present your dissertation.

15.6 REFERENCE

R. Burke, *Project Management: Planning and Control Techniques* (Chichester: Wiley, 2003).

CHAPTER 16

Writing and Presenting the Dissertation

16.1 INTRODUCTION

This chapter is concerned with how we report the results of research. It is not simply a matter of describing what has been done and how it has been done. Good research writing also requires the author to focus on the specific purpose of the research when reporting results and findings. In addition, it must be presented in a style that is easy to understand and that allows the reader to make the connections between the original purpose of the research, its specific objectives and the analysis, conclusions and any recommendations that may arise from it. There is no 'standard' way in which research results are 'written up'; for example, the reporting of medical research is often quite different in style from the social sciences. However, there are a number of 'good practice' elements that we expect to see in a good piece of research writing. This chapter presents many suggestions as to how you should go about designing the layout of your dissertation and what it should contain. These are not comprehensive but they are certainly fundamental to good research writing and good research reporting.

16.2 THE DISSERTATION

This is a vital part of your studies and will contribute substantially to your personal development. You will be expected to demonstrate where appropriate your skills in providing:

- A synthesis of the literature
- An analysis of quantitative and/or qualitative information

- A summary of empirical results whether found by experimentation, observation, survey or interview
- The implications of the findings

Each dissertation should involve some or all of the following:

- Problem identification
- Problem resolution
- Information search
- Application of methods developed in the programme
- Drawing appropriate conclusions

Note: The dissertation is a piece of applied *academic* research and must be more than a mere technical account.

Reporting of research is a vital, but often underestimates facet of research—it is not the most glamorous part but has to be done. Unless you can capture the imagination of others and get them to read your work, all your efforts will have been in vain. You have to think of this as a selling job.

You cannot begin work on writing-up too soon. Ultimately it is how the research will be evaluated. Keeping to the word limit is an important skill, which you need to master. Long, rambling chapters are a sure sign that you have not done enough thoughtful work.

Make use of figures, graphs and schematic diagrams—they are very useful in explaining difficult concepts. However, these must be clearly titled and discussed in the text.

Write in a clear, formal and understandable style. Your work should be understood by an informed but lay audience.

16.3 DISSERTATION OBJECTIVES

There is a set of generic learning outcomes which are expected to be demonstrated in most programmes of study. These are that students should, on completion of the dissertation, have demonstrated:

- An ability to organise and plan their own research activity within the context of their original dissertation specification and time limit.
- The production of a logical, coherent and well-structured analysis of both existing knowledge of their dissertation field and their own contribution to that field.
- The applicability of concepts learned in the taught programme to their specific field of applied research and the critical ability to evaluate the limitations of these as applied to that field.

- The relevance of their work to their organisation generally (where applicable) and to specific issues within the organisation with which they are involved. An example of very typical criteria is provided below:

Typical Criterion	Typical Weighting (%)
Problem formulation: Relevance of the research topic, formulation of the research problem and setting out of the research problem.	10
Research method employed: Validity and effectiveness of the research methods used.	15
Content: Critical appraisal of the literature and evaluation of relevant data.	30
Quality of argument: The extent to which arguments are advanced on valid and reliable evidence. The use of a theoretical framework in advancing themes and ideas.	25
Conclusions and recommendations: Extent to which the dissertation meets its stated objectives. Whether the recommendations are consistent with the evidence and are feasible.	20

16.4 WHAT SHOULD A DISSERTATION LOOK LIKE?

It should reveal all of the following characteristics:

- Nature of the work
- Relevance
- Word limit
- Timetable
- Presentation

Nature of the Work

The main purpose of the dissertation is to enable you to demonstrate to the satisfaction of the examiners that you can undertake an independent piece of research in a specialist area of your choice.

This will involve you in showing that you can design, implement and defend a research project in terms of the research problem identified, the research method(s) used and the conclusions arrived at.

Relevance

Relevancy can be viewed from a number of perspectives:

- The research topic may be heavily policy oriented or heavily theory oriented.
- It may involve a high degree of statistical analysis or a high degree of discursive analysis.
- It may be strongly linked to a single theoretical framework or to a number of theoretical models.
- It may be a work-based research project or not.

For work which is not based on a specific organisation (a single case study) care must be taken that the research topic and applications therein are relevant to a number of organisational situations. This may be done by ensuring that sufficient data is gathered from other organisations. However, where the work is based on a single case study, care must also be taken to ensure that there are generalisations which can be drawn from such work. The main point is that whatever the 'mix' you choose to adopt in your approach to the dissertation it should involve some combination of all four perspectives. That is, it is dangerous to so narrowly define your research topic that its conclusions cannot, in any reasonable way, be capable of at least some generalisation.

16.5 PRESENTING THE DISSERTATION

Presenting a piece of research work is as important as including the type of content discussed in the previous section. Presentation needs to focus on the following:

- Structure
- Aims and objectives
- Current knowledge
- Research method
- Analysis
- Conclusions
- Style

Structure

The structure of the final dissertation is normally presented as follows:

> First inside page = Title, your name and year (2007, 2008, etc.)
> Acknowledgements
> A contents page

List of tables and figures
Abstract
First Chapter = Introduction (research purpose and objectives)
Chapters 2, 3, 4, etc.
Final Chapter = Conclusions
Appendices
References and Bibliography

The abstract should be no more than about 500 words and take no more than one page of space. It should, very briefly, provide a description of the nature of the work, how it was undertaken and its main findings. The dissertation should also contain appendices that include, for example, a copy of a survey questionnaire or interview schedule if these were used in collecting data, a declaration that the data collected will be treated confidentially (see Appendix I to this Chapter), a declaration that the work is yours (see Appendix II to this Chapter) and any other information that you feel should not be in the body of the dissertation but must still be included.

Aims and Objectives

Chapter 1 (introduction) should contain the following elements: the purpose of the research, that is, the research 'problem' (expanding on the abstract) and reasons why it was undertaken. A clear statement of the overall aim and specific objectives of the research should be given. The latter represent the hypotheses or propositions which the research is intended to test.

Research Method

A section on research method and why this method is appropriate to the research questions. Identification of the main data sources used for the research including the limits of validity you have identified in relation to the data. A section on the structure of the text, that is, a brief explanation of what each chapter is about.

Current Knowledge

Chapter 2 should contain your literature review. This should clearly identify where the gaps in knowledge concerning your research topic are relating these to your research objectives. Discussion of the literature should be in a critical mode, which you can use to point to following chapters in which the weaknesses in the literature identified are addressed by your research. In other words, the literature review should be a bridge between the objectives of the research and the analysis to come. An essential stage in any research work is to review the literature; the purpose of this is to know:

1. If the work has already been done
2. Identify the experts in the field
3. Select appropriate research methods
4. Understand where problems can lie
5. Appreciate the debate in the area and where controversies lie
6. Help in generating hypothesis to be tested in the research

A good review should demonstrate familiarity with the topic, show the path of prior research and how it is linked to the current project, integrate and summarise the literature.

To do this effectively the review should be written in a critical and reflective style. One should not simply accept something because it is written; judgement should be passed on it, showing where it is good or where it is poor. Being critical does not mean simply to pick holes in an argument—praise should be given to good ideas.

An example of a good style in literature review would be as follows.

In 1984 Bongaarts et al. published their seminal work on the factors influencing total fertility. This has been greeted with great critical acclaim by demographers such as Smith (1995) and Davies and Ray (1987). However, sociologists, among them Karena (1989), and anthropologists such as Michalson (1987) and Stark (1990) questioned the cultural ethnocentricity of the work. They point out that the ideas of Bongaarts et al. were rooted in a specific culture at a particular time period and as such generalisations to other cultures and times were not reliable. Bongaarts (1998) has now updated his work and linked fertility with a human development index, which relates the model of fertility to economic and cultural development and has, to an extent, satisfied the earlier critics.

Analysis

Chapters 3, 4, etc., should contain the substantive analysis of the research questions identified in Chapter 1 and point to the critique in the literature review where appropriate. This enables the focus of the 'argument' to be maintained throughout.

Conclusions

The Conclusions chapter clearly needs to be relevant to the 'evidence' cited in the substantive analysis. It should clearly show which of the research objectives have been achieved and which remain 'unanswered'. The conclusions should contain a discussion of the 'limits' of the research in terms of: the research method and specific research instruments used, the theoretical framework used, the data analysed and the assumptions made. Additionally, you should be able to point to particular aspects of the research topic which require further investigation.

Style

A number of presentation (format) style rules should be adopted.
Dissertation Title = BOLD, CAPITALS, 18 point
Chapter Titles = bold, Initial Capitals, 14 point
Sub-headings = as Chapters but 12 point
Quotations = quotation marks to be used and quotation to be indented one space below paragraph and one space above the next paragraph. Source and page number(s) to be clearly shown, for example,

> The distinction between equilibrium and disequilibrium measures has become more powerful, with a considerable weakening in the power of disequilibrium measures.
> (Llewellan & Holmes, 1991: 94)

Citing Literature

You must reference all the literature you discuss in the dissertation. This is normally done using the Harvard referencing system. The reason for this is others can refer to your sources, so it must be traceable. The form used for referencing literature is the Harvard system. For web information/literature reference these as: Guerra A.S., Laitana R.F. and Pimpinella M., (1996), Characteristics of the Absorbed Dose to Water Standard. http://www.iop.org/EJ/abstract (accessed 17 July 1998).

Note that authors with two or more publications in a year would appear, for example, Smith (1996a) and then for the other publication Smith (1996b).

Diagrams, charts and tables should be titled and numbered relevant to the chapter in which they appear. That is, Diagram 3.2 is the second diagram in Chapter 3. All illustrations (diagrams, charts, tables) should appear on or close to the text page in which they are discussed. They should not be confined to an appendix. Appendices should only be used for items such as questionnaires, essential extracts, substantial computer output and other data tables which are too detailed for the body of the text.

You are not allowed to use a photocopy of an illustration from an original source without copyright permission.

Typing

Normally your typescript should be one-and-a-half line spaced with a left margin throughout, of at least 25 mm. The typescript should be 12 point and black Arial font. Page numbers should be consecutive and in Arabic numerals.

Initial pages (contents, abstract, etc., and appendices) should be in Roman numerals. All page numbers to be at the bottom centre of the page.

The dissertation must normally be in English and a declaration should be made that the work is the author's own and has not been submitted previously for the award of any other qualification or as a component of any other work undertaken by the author.

The dissertation may be printed on one side of the paper. One-and-a-half line spacing is usually used.

A4 paper of good quality is to be used; margins are normally as follows:

Left (binding edge)   25 mm
Other margins   25 mm

Each chapter should be sectioned into subsections, and the subsections numbered and given a title, for example, section 8 in Chapter 2 would appear as:
2.8 The Value of the Audit

> *Figures* and *tables* These should be included in the main text and referred to by chapter, subsection and number.
>
> For example, referring to a Pie chart the fourth figure in Chapter 3 section 2, might be referred to as displayed in Figure 3.2.4. After the figure a legend should appear, that is   Figure 3.2.4-Pie Chart.
> Likewise for tables.
>
> *References* These should be referred by name and date in the text and listed alphabetically in the bibliography at the end of the dissertation.
>
> For example, the statement may appear:
> 'This finding is supported by Smith (1990) and Jack (1987).'
> This would appear in the bibliography as:
> Jack, B. 'Quality Improvement', Wiley, Chichester, 1987.
> that is, *for a book*: name, date, title, publisher, where published.
> Smith, V.T., 'Improving Quality', International Journal of Quality Improvement, 6, 2, 10–21, 1990.

for a journal: the layout is name, date, title of article, title of journal, volume number, part number, inclusive page numbers.

Binding: It is important that the front and back cover should be resilient to fading. An example of what the front (hardback) cover could look like is given in Appendix III to this chapter and what the first inside page could look like (Appendix IV).

This chapter has given you some key ideas in what a dissertation should contain and how it should be presented. As stated earlier, every student is different in terms of writing style and approach but although your dissertation may not look exactly as described in this chapter it should not look very different!

Appendices

APPENDIX I: CONFIDENTIALITY IN USE OF DATA PROVIDED BY THIRD PARTIES

The data received from the organisations listed below have been used solely in the pursuit of the academic objectives of the work contained in this dissertation and has not and will not be used for any other purpose outwith that agreed to by the provider of the data.

Name (Print): _____

Signature: _____

Date: _____

List of Data Providers

_____

_____

_____

APPENDIX II: DECLARATION

I declare that the work undertaken for this dissertation has been undertaken by myself and the final dissertation produced by me. The work has not been submitted in part or in whole in regard to any other academic qualification.

Title of Dissertation:
_____
_____

Name (Print): _____

Signature: _____

Date: _____

APPENDIX III: SPECIMEN TITLE PAGE (FRONT COVER)

Jupiter University Business School

Bachelor of Science
Bachelor of Arts
Master of Science
in
[Name of Programme]

An Analysis of the Impact of Globalisation
on
Labour Migration

by
A.N. Other
July 2014

APPENDIX IV: SPECIMEN TITLE PAGE (INSIDE PAGE)

> An Analysis of the Impact of Globalisation
> on
> Labour Migration

by
A.N. Other

July 2014

Thesis submitted in partial fulfilment
of the Degree of

Bachelor of Science
Bachelor of Arts
Master of Science
in
[Name of Programme]

APPENDIX V: MULTIPLE-CHOICE SELF-TEST

Now that you have completed this book try to test your knowledge of research methods. There are 20 MC questions below and the answers are given at the end. Try to answer each question—and then look at the answers. If you score between 0 and 5 you need to revise these parts in the book. If you score 10 then you have understood quite a lot, but not enough. If you manage to score up to 15 then you have a good understanding of the concepts. If you can score between 16 and 20 then you have fully understood the key contents of this book. Good luck!

1. A longitudinal research study is concerned with:
    a   A study of two or more research issues at the same time
    b   The period of time a study takes to be completed
    c   Measurements of distance
    d   A study of one or more research issues over a period of time

2. Research Methods and Research Methodology are:
   a   The same thing
   b   Nothing to do with each other
   c   The first comes from the second
   d   The second comes from the first

3. Inductivism is based on:
   a   Numerical data only
   b   Empirical observations of phenomena
   c   Testing the predictions of a theory
   d   Qualitative data only

4. A hypothesis is:
   a   A research question
   b   A research objective
   c   A statement to be proved
   d   A statement to be statistically tested

5. An evaluative, exploratory and instrumental review are all examples of:
   a   Critical reviews
   b   Masters dissertations
   c   Literature reviews
   d   Research studies

6. A cross-sectional research design focuses on:
   a   Relationships between variables over time
   b   A cross-section of the population
   c   Relationships between variables in the past
   d   Relationships between variables now

7. A nominal measurement scale measures:
   a   Categories of things
   b   Ranking of things
   c   Time
   d   Percentages

8. A sample where every unit has an equal chance of being included is:
   a   A stratified sample
   b   A systematic sample
   c   A probability sample
   d   A cluster sample

9. If a research statement is logically true, then:
   a It must also be materially true
   b It may or may not be materially true
   c It will be materially true if its premise is materially true
   d None of the above

10. Deductivism requires that:
    a Enough data is produced to generate predictions
    b Hypotheses are tested against predictions
    c Data is produced to support predictions
    d Any hypothesis is materially true

11. Ordinal data is:
    a Ranked data
    b Numbers
    c Ordinary data
    d Categorical data

12. The purpose of critical reading is to:
    a Be able to find mistakes in research studies
    b Enable the reader to take an opposite view
    c Enable the reader to re-define an argument
    d Enable the reader to evaluate the validity of a research study

13. The measurement of any variable is reliable if:
    a It has been clearly defined
    b Repeated measurement gives the same result
    c Repeated measurement gives nearly the same result
    d The existence of the variable is materially true

14. Primary data is data that has been:
    a Collected by someone else
    b Collected by the government
    c Collected by you
    d Collected from a website

15. The standard deviation of N observations is given by:
    a The square root of the sample size
    b The variance divided by 2
    c The square root of the variance
    d The mean of the sample divided by N

16. A Likert scale is an example of:
    a   A semantic scale
    b   A longitudinal scale
    c   A percentage scale
    d   A categorical scale

17. The coefficient of variation is given by:
    a   Standard deviation divided by the mode
    b   Standard deviation divided by the mean
    c   Variance multiplied by the mean
    d   The mean divided by the standard deviation

18. An odd number scale is commonly used in surveys because it allows for:
    a   A normal distribution
    b   Zero neutrality
    c   A distribution that is not normal
    d   Independence of responses

19. Secondary data is data that has been:
    a   Collected by you
    b   Collected by someone else
    c   Collected in a survey
    d   Collected in a case study

20. A Paired Sample T-test is used to compare:
    a   Samples that are hypothesised to be related in some way
    b   Samples that are hypothesised to be the same
    c   Samples that are hypothesised not to be related to each other
    d   Samples that are hypothesised to be only negatively related

Answers are on the next page

Answers

1. D
2. C
3. B
4. D
5. C
6. D
7. A
8. C
9. B
10. B
11. A
12. D
13. C
14. C
15. C
16. A
17. B
18. A
19. B
20. A

Bibliography and Further Reading

The following list is by no means exhaustive, but it includes a range of texts covering research methods, methodology, statistical techniques and philosophy.

J. Bell, *Doing Your Own Research Project* (Milton Keynes: Open University, 1987).
A. Bryman, *Doing Research in Organisations* (London: Routledge, 1988).
A. Bryman and D. Cramer, *Quantitative Data Analysis with SPSS a Guide for Social Scientists* (London: Routledge, 1997).
G.E.P. Box, W.G. Hunter and J.S. Hunter, *Statistics for Experimenters* (Chichester: Wiley, 1978).
D.R. Cooper and P.S. Schindler, *Business Research Methods*, Seventh Edition (New York: McGraw-Hill, International Edition, 2001).
A.C. Darnell and J. Lynne Evans, *The Limits of Econometrics* (Aldershot: Edward Elgar, 1990).
S. Davies, H. Haltiwanger and S. Schuh. 'Small Business and Job Creation: Dissecting the Myths and Re-Assessing the Facts'. NBER Working Paper, Number 4492, October 1993.
D.A. de Vaus, *Surveys in Social Research*, Third Edition (London: Allen & Unwin, 1991).
G. Dunn, *Design and Analysis of Reliability Studies, the Statistical Evaluation of Measurement Errors* (London: Edward Arnold, 1989).
R. Fletcher, *John Stuart Mill* (London: Nelson, 1971).
U. Flick, *An Introduction to Qualitative Research* (London: SAGE Publications, 1998).
C. Frankfort-Nachmias and D. Nachmias, *Research Methods in the Social Sciences* (London: Arnold, 1996).
D. Futrell, 'Ten Reasons Why Surveys Fail', *Quality Progress* 2 (1994): 65–69.
A.M. Graziano and M.L. Raulin, *Research Methods a Process of Inquiry* (New York: HarperCollins, 1993).
T. Greenfield, *Research Methods: Guidance for Post Graduates* (London: Arnold, 1996).
P. Honey, 'The Repertory Grid in Action', *Industrial and Commercial Training* 11 (1977): 452–459.
D. Hume, *An Inquiry Concerning Human Understanding* (London: Bobbs-Meril Publishers, 1748).
T. Khun, *The Structure of Scientific Revolutions* (Second edition) (Chicago: University of Chicago Press, 1970).
G. Lang and G.D. Heiss, *A Practical Guide to Research Methods* (London: Lanhan, 1984).
D.C. Miller, *Handbook of Research Design and Social Measurement* (London: SAGE Publications, 1991).
K. Miller, 'Are Your Surveys Only Suitable for Wrapping Fish?', *Quality Progress* 4 (1998): 47–51.
D.C. Montgomery, *Design and Analysis of Experiments* (Chichester: Wiley, 1984).
D.L. Morgan, *The Focus group Guidebook* (London: SAGE Publilcations, 1998).
R.L. Ott, *An Introduction to Statistical Methods and Data Analysis* (California: Duxbury Press, 1988).
W.D. Perrault and L.E. Leigh, 'Reliability of Nominal Data Based on Qualitative Judgements', *Journal of Marketing Research* XXVI (1989): 135–148.
H. Poincare, *Science and Hypothesis* (English translation 1905) (New York: Science Press, 1902).
K. Popper, *The Logic of Scientific Discovery* (Second edition) (London: Hutchison, 1968).
C. Robson, *Real World Research* (Oxford: Blackwell Science, 1995).
Harry F. Wolcott. *Writing Up Qualitative Research* (SAGE Publications, Inc, 1990) http://www.sagepub.com/books/Book3147
R.K. Yin, *Case Study Research, Design and Methods* (London: SAGE Publications, 1994).

About the Authors

John Adams is the Head of Department of Economics at the British University in Egypt and was previously Director of the China-EU Research Centre based in Edinburgh. He has published widely in the field of economics in both national and international academic journals, is the co-author of several textbooks and has worked as external examiner at the Universities of Dundee, Sussex and Hong Kong. Professor Adams has also undertaken consultancy assignments for a wide range of public and private sector organisations in the United Kingdom and overseas. He is currently a visiting Professor at three universities in China. His research is mainly focused on economic development and on the process of development. Professor Adams also provides training to Investment Promotion Agencies in several countries on attracting, managing and evaluating FDI.

Hafiz T. A. Khan is a Senior Lecturer in Applied Statistics in the Department of Economics and International Development and is a Demographer at the Centre for Research into the Older Workforce (CROW) in the Middlesex University Business School. He is also visiting Research Fellow in Demography at the Oxford Institute of Population Ageing, The University of Oxford. Dr Khan trained as a statistician at the University of Chittagong and later as a demographer at several institutions: Edinburgh Napier University, the International Institute for Applied Systems Analysis (IIASA) in Austria, the National University of Singapore and lastly at the University of Oxford. Dr Khan's principal research interests lie in the broader areas of population and development including population ageing and its consequences, poverty and vulnerability, microfinance; development issues, reproductive health and family planning in developing countries. For a number of years he has also worked on the demographic issues of Bangladesh especially on the trends, determinants and differentials of fertility as well as elderly care and support. He has written extensively in population-related issues and has over 90 academic publications including books, book chapters and journals.

Robert Raeside is currently Professor of Applied Statistics and Director of the Employment Research Institute (ERI) at Edinburgh Napier University, UK. He is also a Chartered Statistician of the Royal Statistical Society (CStat), London, Member of the Operational Research Society, Birmingham, and a Fellow of the Higher Education Academy (FHEA), York.

He received his BSc Hons from Napier College under the auspices of the Council for Academic Awards (CNAA), his MSc from the University of Strathclyde, Glasgow, and PhD in demographic forecasting from the CNAA in collaboration with the University of Oxford. He serves on two British Standards Committees and is part of the editorial teams for the *Journal of Applied Probability and Statistics* and the *Journal of Revenue and Pricing Management*. He was appointed a visiting Professor in Demography to the University of Dhaka, Bangladesh, 2009.

Professor Raeside's research has been on the application of statistics to areas of demographic change, public health, employment and business improvement. He is particularly interested in the use of social network analysis to investigate complex processes. He has acted in an advisory capacity to business, local authorities and to the national government.